# THE INDUSTRIAL DESIGN REFERENCE + SPECIFICATION BOOK

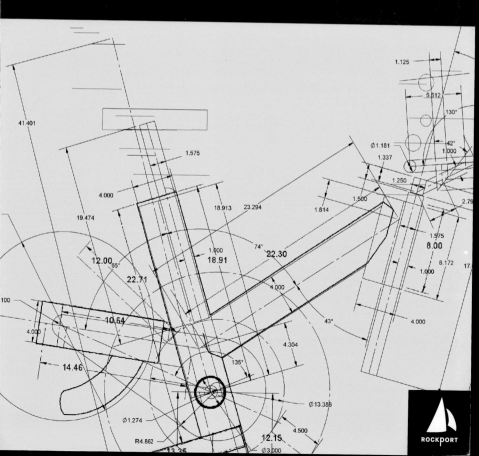

ROCKPORT

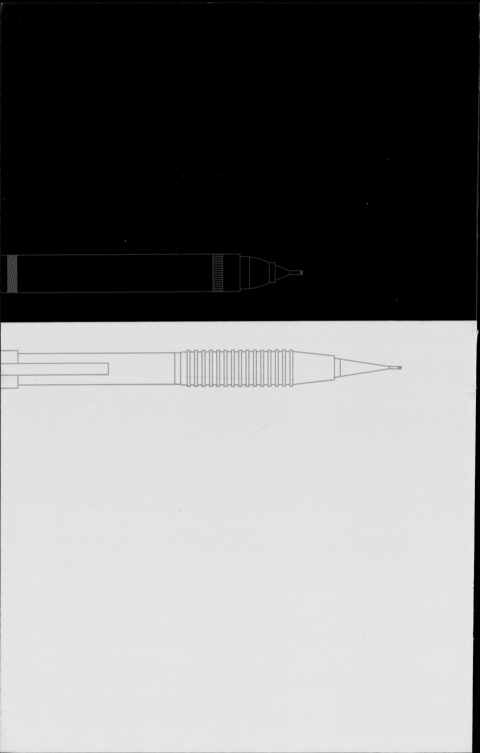

# THE INDUSTRIAL DESIGN REFERENCE + SPECIFICATION BOOK

ALL THE DETAILS INDUSTRIAL

DESIGNERS NEED TO KNOW

BUT CAN NEVER FIND

Daniel F. Cuffaro

Douglas Paige

Carla J. Blackman

David Laituri

Darrell E. Covert

Lawrence M. Sears

Amy Nehez-Cuffaro

Updated materials provided by

Issac Zaks

© 2006, 2013 Rockport Publishers

First published in the United States of America in 2013 by
Rockport Publishers, a member of
Quarto Publishing Group USA Inc.
100 Cummings Center
Suite 406-L
Beverly, Massachusetts 01915-6101
Telephone: (978) 282-9590
Fax: (978) 283-2742
www.rockpub.com
Visit RockPaperInk.com to share your opinions, creations, and passion for design.

Originally found under the following Library of Congress Cataloging-in-Publication Data
  Cuffaro, Daniel
  Process, materials, and measurements : all the details industrial designers need to know
  but can never find / Daniel Cuffaro.
    p. cm.
  Includes bibliographical references and index.
  ISBN 1-59253-221-7 (hardback : alk. paper)
  1. Design, Industrial. I. Title.
  TS171.C84 2006
  745.2—dc22                                                  2005030689
                                                                    CIP

ISBN: 978-1-59253-847-8

Digital edition published in 2013

eISBN: 978-1-61058-789-1

10 9 8 7 6 5 4

Design: Peter King & Company
Layout and Production: Leslie Haimes
Cover Design: Burge Agency, www.burgeagency.com

Printed in China

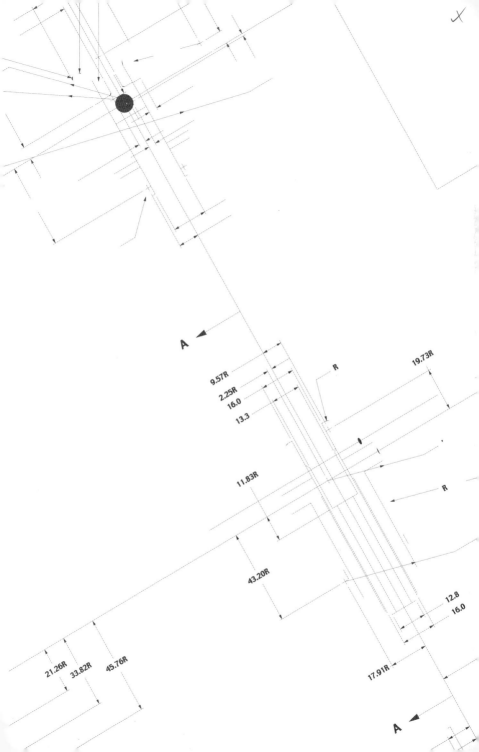

# ○ Contents

# Introduction

Industrial designers involved in developing a product often refer to many sources for measurement conversions, manufacturing processes, material selection, or product development terminology. Rarely can industrial designers find a common and concise source of important information. In many cases, the sources of necessary information for designers delve into great depth about details that have little effect on the final product.

*The Industrial Design Reference & Specification Book* is a compilation of essential information that intends to function as a reference manual for the things a designer needs to know on a daily basis. This book includes a wide range of information, from understanding the larger context of business and process to material selection. In addition, important information on intellectual property, ergonomics, and documentation is included.

*The Industrial Design Reference & Specification Book* serves to answer the questions that designers ask most often and deliver the answer in a clear and concise way.

⌀18.00

—**Daniel F. Cuffaro**

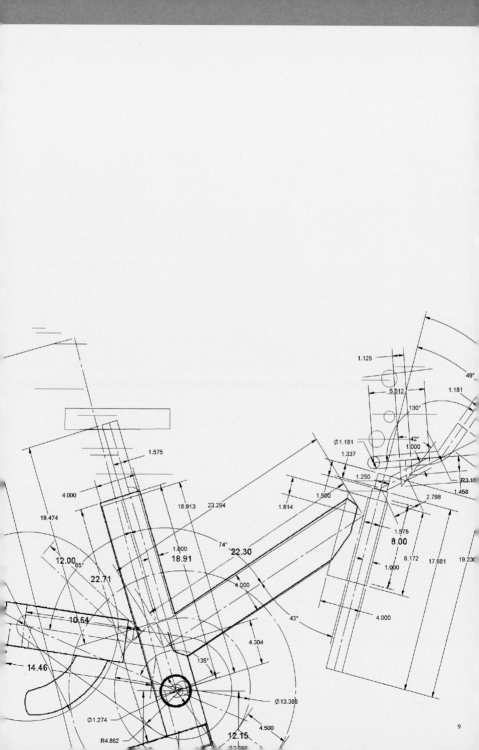

# Chapter 1: Design in the Context of Business

by David Laituri

## OVERVIEW

While many chapters in this book concentrate on various aspects of the day-to-day design process, this chapter provides a basic, abbreviated overview of the larger business environment in which design exists (and would not exist without), as well as some of the ways design interacts within this larger environment. Design is a fundamental business function, as are finance, logistics, operations, marketing, sales, and research and development, depending on the company. For a business to be successful, and to maximize its resources, all of these basic functional units must work well together.

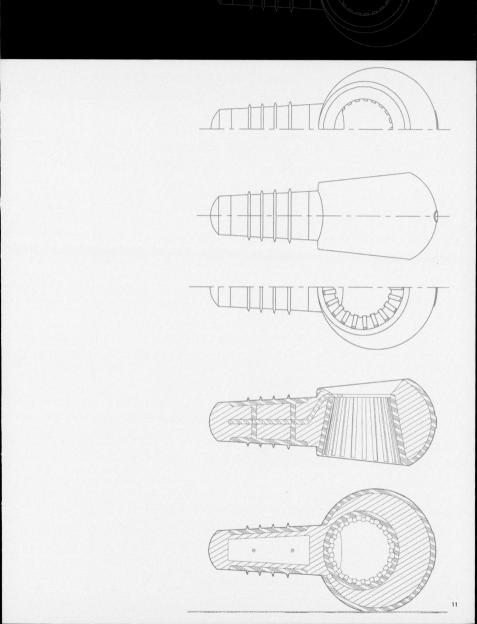

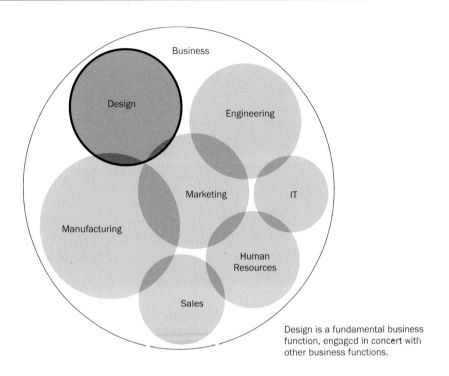

Design is a fundamental business function, engaged in concert with other business functions.

Design in all its forms was born within a business environment, to meet a business need or serve a business purpose. The design/business relationship, some argue, is what separates the applied arts, particularly design, from the fine arts, such as painting, photography, or sculpture. The "make more than one" dimension of design—whether for print, commerce, or mass production—is what inextricably ties design to business. Design, at its most basic level, is about helping businesses seek a sustainable advantage in their given market, an advantage that leads to consumer preference and, in turn, to improved profits.

In the early stages of your design career, it is important to realize that as you move forward in this profession, you will be required to understand these basic concepts and apply them in your daily professional life. As design continues to evolve and mature as a critical competitive advantage, it is safe to assume that business will play more of a role in design's future, not less.

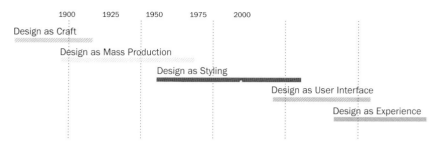

The evolution of the design profession has mirrored development of the surrounding economy.

## THE BUSINESS OF DESIGN

Businesses share a fundamental purpose, which is to develop products and/or services to exchange with consumers for a profit. To accomplish this, a business must identify a product, a service, or some combination, that targets a consumer's need or desire; and bring together the financial, raw material, and labor resources to develop and deliver this product and/or service efficiently to these consumers. The two key concepts to note here are efficiency, the continual effort to maximize use of financial, labor, and other resources; and effectiveness, the continual effort to maximize purchases by the consumer, thereby assuring the highest possible retained earnings, or profits. This profit-creation objective is arguably the central element of all business activity. Without profits, a company would eventually deplete its financial resources and cease to exist. Ignoring the central role of profit in any business is detrimental, and businesses' investments in design functions are no exception.

## MARKETING

Planning the most efficient and effective path to this profitable contribution is known as developing a design strategy. The design strategy needs to account for the company's product sector, market segment, various strengths and weaknesses, technological assets, manufacturing capabilities, channels of distribution, and competitive products or services in the market. It also needs to understand the target customer's attitudes, needs, and desires. Every product sector has a different mix of competitive forces at play; even two different companies in the same sector will approach the same situation differently. The design strategy needs to identify the most efficient and effective way to use a company's financial and human resources to achieve success in the market and add value to the brand.

Businesses divide into two basic categories based on their financing: private companies operate with funds they generate or borrow, and public companies sell all or part of themselves to shareholders. These shareholders can be individuals, banks, or investment groups who purchase and effectively own portions of the company called shares. As the company's sales performance improves or declines, so does the perceived value of these shares. Shares of a company are, in turn, bought and sold in exchanges such as the New York Stock Exchange and the Tokyo Stock Exchange. These publicly held companies are subject to specific rules and regulations designed to ensure ethical conduct and protection of shareholder's interests. Going public with an Initial Public Offering (IPO) is a common method used by private companies to raise funds for their capital-intensive ventures, ones that require large amounts of cash to develop a particular product or service. Without the ability for companies to sell their own shares and raise necessary capital, most products and services today would not exist.

With profit-creation being the primary objective of any business, public or private, it is the role of the business leader—the Chief Executive Officer (CEO), in most cases—to seek the maximum return on their shareholder's financial investment. This return is measured in the increase in value of the company's shares based on the real or perceived success of the company has in selling its product or service or in reducing costs of operation. The company's shareholders hire the CEO through a group of independent, often external advisors, which they elect to serve on the Board of Directors. The Board of Directors continually reviews the state of the business and advises the CEO on key decisions that might affect the Return on Investment (ROI) of the shareholders who elected them. If the Board of Directors determines that a CEO is not making the most of their financial investment, they are empowered to seek a new leader that can; if he exceeds their expectations, they can reward him financially through bonuses, shares of stock and other incentives as well. This carrot-and-stick combination is a powerful tool for the Board to direct the CEO to, in turn, enlist all direct-reporting managers to ensure that all of their individual departments' activities are fully aligned to the mission given him by shareholders through the Board.

Prior to the beginning of a company's financial year, or fiscal year, the Board and CEO develop a target financial goal to achieve in the next twelve-month period. This goal takes into account elements such as the costs of developing new products, opening new retail outlets, the state of the surrounding economy, and the potential actions of key competitors, to name a few. Once this goal and the strategy for achieving it are in place, all actions of the company, the CEO, and the officers will be measured continually against the goal. The departments of the business are, in turn, given similar goals that, when enacted in concert with each other, contribute toward achieving the agreed-upon business strategy for the year. To achieve their individual goals, each function of the business must develop a strategy that works in concert with other departments to contribute to the overarching strategy of the business.

## STRATEGY

The term strategy is a very broad business concept that could easily fill a chapter of its own. There are countless volumes written about corporate and business strategy, as well as dedicated fields of academic pursuit. Strategy can simultaneously refer to a plan, a process, or an outcome. For our purpose here, strategy is defined as an attempt by a company to arrange its resources (labor, money, raw materials, factories, etc.) for maximum competitive advantage prior to engaging them. Strategies are essentially plans for the future of a business. Tactics, by contrast, are all of the actions taken to realize that strategy (developing new products, building new factories, setting prices, hiring more people, reducing costs, etc.). Strategies can be long term, such as entering a new product area or selling a current product line in a new country, or they can be short term, such as temporarily cutting prices to gain market share quickly from a competitor.

There are two basic positions that companies will take relative to their markets. The first is a pull position, where the market may have many mature competitors; the product or service may seem to be interchangeable with competitors'; and brand loyalty among consumers may be low. In this case, it is critical that design and marketing invest considerable effort in understanding users and their subtle perceptions of the product, service, or brand, as well as their sometimes hidden unmet needs. A company's success or failure in crowded commodity markets often hinges on finding

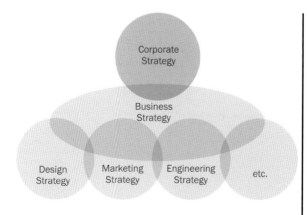

**Focuses on:**

**Corporate Level**
- Financial goal setting
- Mergers/Acquisitions
- Shareholder relations

**Business Level**
- Which markets?
- Which technologies?
- Product planning

**Functional Level:**
- Tactics of individual functions

**Strategy Pyramid:** the strategies developed and employed by various functions support the strategy of the business. Effective strategy setting works in reverse; a vision for the business must be in place first, and the steps (strategy) for achieving that vision work backwards from it.

ways to understand and exploit these unmet needs as quickly as possible. General Motors and Proctor & Gamble are examples of companies who rely heavily on consumer insights to drive product development.

The second position companies might take is a push position. Whether a market is saturated or not, a company may choose to invent a wholly new product or service that solves a particular problem or achieves a particular level of quality, then communicate to consumers the benefits of this invention. Inventing wholly new solutions is both expensive and risky. Inventions do not happen on a predictable schedule and it often takes many attempts to find one that is successful, and the company's financial and time investment may not generate the desired return in the market. However, if a company does create an invention that consumers identify with, the financial rewards can be considerable. Apple Computer and Bose are examples of companies who innovate first and then lead consumers to their improved product solution through communications, direct mail, and advertising.

## COMPETITIVE FORCES

All businesses face some form of competition. Competitors can take the form of direct rivals, who offer a similar product or service, as well as competitive forces, which take the form of indirect competitors. Direct competition can be easier to identify and develop a strategy against; note the decades-old competition between Ford, Chrysler, and General Motors in the U.S. auto industry. Competitive forces, by contrast, can be difficult to plan for, such as unforeseen increases in raw material prices, an unexpected drop in consumer confidence, or increased regulation by the Federal government.

Seek needs first or invent solutions first? The two basic strategies companies can use have strengths and weaknesses, but both can be effective.

As design and business have evolved together over the years, competitors have sought to emulate advantages gained by businesses that understood the role of design in developing consumer preference for a particular product or service. As one company in a product sector began to use design to differentiate its products and services, another would copy these efforts, ultimately eroding the initial advantage. While every area of a business experiences a competitive arms race, this pressure has caused design to evolve beyond the creation of superficial "looks" for products, which are easy to copy. Now, it is more common to use a targeted design strategy, which is more difficult to copy.

There are several models that describe the competitive environment of a business, one of the most well known being that of Harvard Economist Michael Porter. Porter's Five Forces model lays out the basic external forces that apply generically to any industry: the power of suppliers, distributors, new entrants, substitutes, and direct competitors. Understanding these forces is a critical first step in determining and maintaining an effective, profitable business strategy.

Since one of the basic functions of any business is to take inputs (i.e., raw materials, energy, money, components) and add value to them, suppliers of these inputs play an important role. Suppliers are powerful when there are few of them for a particular input. In the case of a factory, there is often only one supplier of electricity, making the electric company relatively powerful. Suppliers are weak when there are many of them—for the same factory, labor may be plentiful and not organized in the form of trade union, making the supplier of labor relatively weak.

Distributors, or channels, for a particular product or service are any entity from which the end user purchases the product or service. Like suppliers, channels can be powerful if few in number, or when the channels represent a large share of potential buyers. Large, mass-market retailers such as Wal-Mart or Home Depot in the U.S.

Michael Porter's Five Forces model is a useful tool for examining the external forces on a business in any given market: essential information for setting effective strategies for a business.

are examples of powerful distributors. Mass-market retail outlets often account for a disproportionate percentage of sales for a company, making them relatively powerful when compared to other retail options. Again, like suppliers, the power of distributors is relatively weak when there are many, seemingly interchangeable options for consumers, such as car rental agencies at the airport of a major city.

New entrants include anyone who might view a particular product sector as financially lucrative enough to create a competitive product or service for sale and distribution. Existing companies in a particular product area can create barriers to keep new entrants out of a market by maintaining high manufacturing costs, or through developing high brand preference among consumers. The commercial aircraft industry is an example of one with very high manufacturing barriers, or costs of entry. The athletic shoe industry is an example of one with high brand preference barriers. Conversely, restaurants and auto repair businesses in any given town are examples of ones with relatively low barriers to new potential entrants.

As mentioned before, competitors in the Five Forces model are those who make a similar product or service—one that consumers might consider interchangeable on some level. Direct competitors are one of the most obvious market forces, as their products are the ones adjacent on the same store shelf, often with similar features at a similar price. Direct competition is most obvious in a typical grocery store. The force of direct competitors is weaker when the company has a unique product protected by patents, trade dress, or copyright protection.

Substitutes are any other competing priority for the same consumer dollar. Walking or riding a bicycle is a substitute for using a car and purchasing gas. Eating dinner at home is a substitute for going out to a restaurant. Not spending the money, investing it elsewhere, or simply keeping it in the bank is also an option for consumers. No company's product or service is without some form of substitute.

## BRAND BASICS

Brands are as old as business itself, dating back to a time when two merchants in the same area who provided the same goods or services needed a way to differentiate themselves from each other. Since literacy levels were very low, competitive proprietors used unique symbols to identify their place of business, giving birth to the logo and its associated brand. Creating and communicating differentiation, and building preference for that differentiation, are basic goals in brand building.

While it's easy to identify a brand by its outward-facing visible elements—such as a unique logo, the consistent use of a particular color, or a signature product—

strategic marketers actually manage brands at a much deeper level. They see brands as a way to own an important and compelling idea in consumers' minds, to build loyalty, repeat purchases, and achieve premium prices relative to competitors. Brands represent a promise of quality or function consistently made—and repeatedly kept—to a loyal consumer segment. An example of an effective brand idea would be "Different" for Apple Computer or "Driving Excitement" for BMW. Over time, a consistent collection of brand elements can become synonymous with the compelling idea that the company wishes to own in their consumer's mind.

As products and services have become more alike and similar in price, the subject of brands (identifying and building a positive reputation in consumers minds) and branding (identifying and applying a consistent look, feel and voice to everything a company does) have taken on increased importance in business in recent years. Brands themselves have also become valuable properties, capable of being sold for sums that are

## BRAND ANATOMY

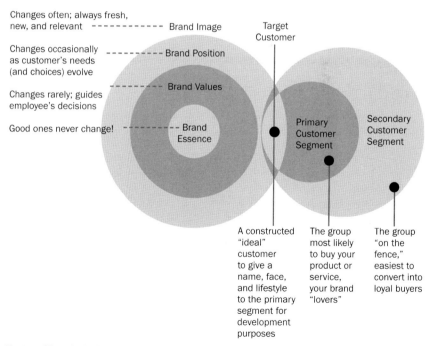

Changes often; always fresh, new, and relevant -- Brand Image

Changes occasionally as customer's needs (and choices) evolve -- Brand Position

Changes rarely; guides employee's decisions -- Brand Values

Good ones never change! -- Brand Essence

Target Customer

Primary Customer Segment

Secondary Customer Segment

A constructed "ideal" customer to give a name, face, and lifestyle to the primary segment for development purposes

The group most likely to buy your product or service, your brand "lovers"

The group "on the fence," easiest to convert into loyal buyers

Illustrated here is the basic anatomy of a brand. While brands "live" where the company and its products or services meet the intended customer segment, fifty percent of a brand's job is to inspire and direct employees—the "holder" of the company's brand.

multiples of the value of the underlying businesses. Building, growing, maintaining, and protecting a brand's value has become a vital responsibility; a mismanaged brand reputation can sometimes kill a business, taking equity and value to shareholders with it.

While designers and marketers have a common interest in brand management, they often view, define, and interact with a brand differently. Marketers see brands as a business; designers see brands as an image. While both definitions are essentially correct, they are incomplete when viewed in isolation. Brand management is the responsibility of anyone who has a role in representing the brand to the world, either through direct contact with consumers or through creation of its products, services, and communications.

## BASIC BRAND CONCEPTS

**Brand essence** A brand's essence is the basic and most fundamental idea at the center of a brand. Some companies start with this idea in mind; others identify it after years of experience in the market. Well-managed brands use their essence to guide and inspire employees as much as they do to build loyalty and preference with their customers.

**Brand values** These are basic guiding principals or rules that organizations use to assure that all employees continue to support and build on a brand's essence and increase its value.

**Brand guidelines** These are documents in a variety of forms created to assure the consistent understanding and use of important brand assets. They are typically shared with new employees as well as external brand execution partners (external design firms, printers, signage companies, etc). Guidelines detail correct as well as incorrect uses of brand elements.

**Brand position** This is a statement marketers develop that captures the specific idea they wish to "own" in their customer's minds above all other competitors. Having a clear position in place helps marketers focus a product or service.

**Brand platform** A term often used by a single brand that may have several different customer targets, with associated products or services. A platform allows a single company to tune their brand position and message in multiple ways to resonate with specific, often differing customer segments.

**Sub Brand** A term used to identify a brand "space" within a larger brand. United Airlines' short-haul commuter business, "Ted," is a good example of a sub brand.

**Brand image** This refers to the combined look, feel, and "impression" of all externally facing brand elements (i.e. the signage, retail environments, advertising tone and manner, packaging, products, websites, and customer service).

**Brand strategy** A plan, usually beginning with a desired future state for a brand, which outlines a variety of tactical steps a brand must undergo to reach its full and desired potential. Brand strategy by necessity is "live" and under continuous review to account for changes in the external market and consumer response.

**Brand manager** The definition of this role can vary widely, depending on the company. It can be an individual charged with running a business, or an individual who actively manages the brand image of a business.

**Brand equity** To a marketer, this refers to the good will a customer has for a company, which can be appealed to for loyal, repeat purchases. To designers, equity can refer to the specific visual elements that customers identify with the company with that also build loyalty and foster repeat purchases.

**Brand assets** Logos, colors, graphic treatments, shapes, forms, and even sounds that a company has invested in, and that consumers have learned over time to associate positively with a particular brand. Mickey Mouse is an important brand asset for Disney, the golden arches are synonymous with McDonald's, and T-Mobile has a trademark sound used in commercials.

**Brand recognition** A metric often used to test consumers on their ability to recall from a name or logo alone a company's product, service, or position. Brand recognition percentages are considered a reflection of the effectiveness of a company's communication efforts.

**Brand preference** The goal of a company is to develop a brand that consumers will prefer over other brands which offer similar products or services.

**Brand loyalty** The degree to which consumers will identify a brand they prefer through trial, ultimately selecting one for continued repeat purchases.

## CONCLUSION

At some point in your professional future, it may make practical sense for you to consider pursuing additional education in business management to improve your understanding and application of these important concepts. Adding advanced knowledge of business is beneficial to anyone, regardless of his or her original professional discipline; designers are no exception. Gaining advanced business credentials, possibly in the form of an MBA (Masters of Business Administration degree), will ultimately help make you a more effective leader of an important and essential business discipline.

# Chapter 2: Process

**by Daniel F. Cuffaro**

## OVERVIEW

Developing a product requires significant human resources and capital, and success is not guaranteed. Many forces affect how well a product performs in the marketplace. To reduce unnecessary risk and present the greatest likelihood for success, it is critical for the development team to establish and practice sound development processes. This ensures that the team possesses information, equipment, and structure to make good decisions and move forward effectively. The process described in this chapter primarily deals with the development of a consumer product, but many of the tasks and involved disciplines are similar in the development of commercial and medical products—although the scale and complexity of the product, and processes, may vary dramatically.

For most industrial designers, the fundamental process has three components: research, conceptualization, and refinement. The designer must inform himself, explore broadly, and make adjustments for manufacturing. While there are many permutations of this process, and many activities are not clear cut or linear, these three stages describe the most simplified representation of the designer's activities in product development.

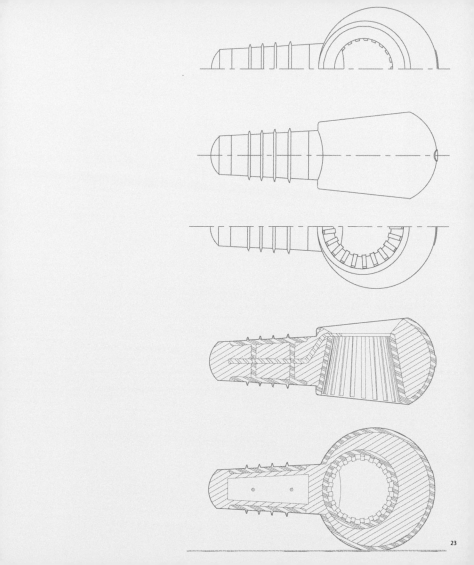

The responsibilities of the design team represent a small set of activities that happen within a much larger context. This chapter intends to illuminate the entire product development process, illustrate the roles of marketers and engineers, and map the relationships and responsibilities of an interdisciplinary team.

## MACRO PROCESS

Within the process of developing a product from beginning to end, there are three primary stages: planning, development, and production. These stages define the process beyond the activities of any specific discipline. A core multidisciplinary team, supplemented by specialists, will work in various degrees and capacities throughout the project. Typically, marketing is said to "own" the responsibility of getting the product to market and therefore has a consistently high level of activity throughout the entire process.

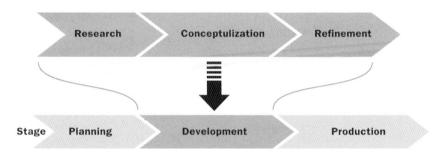

The familiar Research/Conceptualization/Refinement process is a subcomponent of the larger development process.

## PLANNING

The planning stage typically involves an initiating event, such as the need to replace or update an existing product; the identification of a new need or opportunity, or an opportunity to extend an existing product line. Once the opportunity for a new product is identified, market research is conducted, to verify the need, understand the competitive landscape, identify the target market, and understand the overall financial opportunity. Researchers accumulate information through focus groups, surveys, and one-on-one interviews. The data gives the marketing team statistical data that captures demographic information, purchase intent, product preference, and other important information.

| Stage | Planning | | Development | | Production | |
|-------|----------|--|-------------|--|------------|--|
| Activity | Identify Opportunity | Research/Project Definition | Conceptualization | Refinement | Tooling | Ramp-Up |

The product development process can be divided into stages of focused activities, which include planning, development, and production.

Combining the market research and past histories of success and failure creates a preliminary product specification. This specification intends to set goals for cost and performance requirements, as well as define important parameters such as manufacturing processes and package size. This specification functions as a tool for measuring success in the development stage.

Throughout the planning stage, the development team is identifying project and product limitations, and defining the basic structure of a project. Goals are set, a team assembled, and clear project objectives defined.

| Product | Customer Stakeholder | User | Other Stakeholder |
|---------|---------------------|------|-------------------|
| Barcode Scanner | • purchasing agent<br>• IT services<br>• store manager | • checkout clerk<br>• inventory specialist<br>• registry customer | • service and repair<br>• tech support<br>• sales force |
| Mobile Phone | • average consumer<br>• service provider<br>• retail outlet | • average consumer | • service and repair<br>• tech support<br>• sales force |
| Baby Rattle | • gift giver<br>• parent<br>• retail outlet | • baby<br>• parent<br>• siblings | • sales force |

It is important to differentiate the needs of the customer, who actually purchases the product, and the user, who uses the product. In many cases, these individuals are not the same.

**DEVELOPMENT** After the planning stage is completed, the team begins to focus on developing a product that will meet customer and user needs. Industrial designers are most involved in this stage of a project, and the familiar "research, conceptualization, and refinement" activities are paramount. Many of the activities in this phase rely on interactions with marketers, research specialists, and engineers who can provide data, technical support, and process guidance, to assist decision making.

Design research functions as an extension of market research. However, the goal is not to accumulate large amounts of quantitative information; rather the emphasis is on building a dialogue with the customer and user to gain a better understanding of priorities and perceptions. Observing and experiencing use scenarios are critical components of the research, enabling designers to develop an understanding of current issues and opportunities. Typically, data are captured using photos, notes, diagrams, and the like, which primarily functions to provide inspiration for concept generation and create context for the concept presentation. Designers may also research seemingly unrelated activities and products for transferring effective knowledge and techniques from one application to another. After research is completed, a design brief is created, which defines the goals for the look and feel of a product and how it emotionally connects with the customer/user. The product specifications may also be refined based on the design research.

Guided by marketing and design research, conceptualization begins. Working in an interdisciplinary fashion, designers explore product architecture, aesthetics, ergonomics, functionality, technology application, manufacturing, and the like through drawings and mock-ups. Depending on the nature of the project and the time available, concepts may be developed within tightly defined parameters, or designers may have the freedom to broadly explore alternative approaches (and at times suspend reality and restrictions for the sake of inspiring innovative ideas).

At this stage, concept development is typically divergent, and capturing a variety of unique approaches ensures that many aspects of a problem are considered. Solutions are measured relative to the design brief and product specifications, and decisions are made on the direction(s) for refinement. At times, the concept development phase may influence changes to the brief and specification.

The refinement process typically involves a convergence on promising concepts and attributes, which combine to form new refined solutions. The integrated solution takes into account aesthetics, function, material selection, and manufacturing, and focuses on meeting the previously defined specifications and project goals. In addition, suppliers are identified, assembly scenarios are planned, and cost targets are set. During this activity, designers and engineers are working closely together to ensure that a viable direction is well documented and prepared for the production hand-off. The design deliverables typically include detailed renderings, drawing packages, appearance models, and a detailed CAD database. Parallel to the design and engineering activities, the marketing team begins to develop a marketing plan and determine the pricing structure.

**PRODUCTION** After capturing the design intent in the development phase, the team begins to focus on final refinement, tooling development, and ramp-up for mass production. The role of industrial designers begins to diminish in this stage, with the exception of continuing to maintain the design and interface intent, or make appropriate compromises. Design engineers begin to work with manufacturing engineers, detailing the parts and continuing to make refinements with an emphasis on efficient and cost-effective manufacturing. Mechanisms are designed, modeled, and refined, individual parts are evaluated, and the assembly strategy is defined. Parallel activities such as electrical engineering, component sourcing, and testing continue to inform the process and affect refinement. Alpha prototypes are built, which represent the design and functionality of the final product, but not necessarily the identical manufacturing processes. The alpha prototypes are tested and refined as needed. After the creation and refinement of the fully detailed assembly, the team prepares for a tooling release, which comprises drawing packages, CAD databases, bills of materials (BOM), and samples. At this stage, marketing is working with graphic design, packaging, and advertising teams to develop product launch materials, packaging, and product instructions.

After the release, tools are built, and long-lead time items are ordered. Often, tooling is made by one manufacturer and used by another (a toolmaker cuts the tooling, while a molder molds the parts). As the tooling is developed, it is regularly tested and modified (debugged) to ensure that it can effectively produce parts that represent the design intent, the intended quality, and dimensional accuracy. Beta prototypes are built with parts that represent the actual manufacturing process and components. These prototypes are tested extensively to evaluate reliability and performance. In addition, final assembly strategies are determined and parts are refined as needed. Fully functional prototypes are used to gain agency approvals and early customer feedback, and are then refined as necessary.

When it is determined that the product is ready for manufacturing, the production **ramp-up** begins. Assembly lines are set up, the work force is trained, and pre-production products are produced and evaluated. In the initial stages of production, inventory will often be accumulated and shipped to key customers. Adjustment of the production capacity occurs to meet demands, and the process will undergo ongoing evaluation and refinement. Periodic checks of components for quality and ongoing design and process changes are likely. Concurrently, marketing is orchestrating the formal launch, evaluating consumer reaction, and working with design and manufacturing to make adjustments as needed.

After the launch, management of the product lifecycle must continue. Cost reductions occur, color and graphics may change, and service calls must be addressed, all while continuing to track sales performance and customer feedback. Data throughout the entire process then becomes a resource for the planning stage of the next project.

## CONCLUSION

Developing a product is a complex and resource-intensive process that is difficult to map and often unpredictable. Having a good idea and planning well is critical to minimizing unnecessary risk and presenting the greatest opportunity for success. By understanding the goals, stages, relationship of activities and disciplines and by providing appropriate resources, a great product can become a successful product.

# PROCESS DETAIL

| | Overall project blueprint | | Creation of product design | | Manufacturing of product | |
|---|---|---|---|---|---|---|
| **STAGE** | **Planning** | | **Development** | | **Production** | |
| **ACTIVITY** | Identify Opportunity | Research/Project Definition | Conceptualization | Refinement | Tooling | Ramp-up |
| **TASKS** | Tasks in the early planning stages of a project:<br>• Identify need/opportunity<br><br>• Define target market<br><br>• Define business objectives<br><br>• Identify limitations<br><br>• Identify outside forces<br><br>• Determine the project time line<br><br>• Set project goals<br><br>• Define project team<br><br>• Consider prior project learning | • Conduct market research<br><br>• Identify customer<br><br>• Identify user<br><br>• Conduct competitive market analysis<br><br>• Validate need/opportunity, target market, etc.<br><br>• Assess product options<br><br>• Create preliminary specification<br><br>Late activities (in development stage):<br>• Conduct design research<br><br>• Understand customer<br><br>• Understand user<br><br>• Create product specification<br><br>• Create design brief | • Explore aesthetic, ergonomic, functionality, and manufacturing options<br><br>• Create concepts in the context of research (rationale of solution/selection)<br><br>• Build prototypes<br><br>• Develop concept "feedback loop" (user testing)<br><br>• Test concept feasibility<br><br>• Explore patentability<br><br>• Evaluate design/engineering<br><br>• Estimate preliminary costs | Early activities (development stage):<br>• Refine design concepts<br><br>• Conduct preliminary manufacturing analysis<br><br>• Estimate manufacturing cost<br><br>• Create preliminary database(s)<br><br>• Build and test prototypes<br><br>• Conduct user testing<br><br>Later activities:<br>• Handoff from industrial design to engineering team | Engineering activities:<br>• Design tooling<br><br>• Procure tooling<br><br>• Debug tooling<br><br>• Test beta prototype<br><br>Marketing activities:<br>• Finalize marketing plan<br><br>• Develop marketing materials<br><br>• Develop packaging, graphics, and instructions | Early production:<br>• Set up assembly line<br>• Train workforce<br>• Evaluate production output<br>• Begin production<br>• Launch marketing/advertising<br><br>• Fill key customer orders<br><br>Ongoing activities:<br>• Adjust production/inventory relative to demand<br><br>• Monitor customer feedback<br><br>• Monitor quality<br><br>• Explore cost reduction<br><br>• Evaluate design for improvement |

## PROCESS PARTICIPATION LEVEL

| | Planning | Development | Production |
|---|---|---|---|
| Marketing | | | |
| Design | | | |
| Engineering | | | |

# Chapter 3: Manufacturing Processes

by Daniel F. Cuffaro

## OVERVIEW

The selection of a manufacturing process directly relates to the needs of the user, the resources of the producer, and the time available to bring the product to market. The level of sophistication in mass production has increased dramatically, relative to the cost, so that many more processes may be realistically considered. In the field of industrial design, plastics have become the primary medium, with metals being a close second. The importance of understanding the forming, finishing, and assembly of these materials is critical. In addition, designers are now more likely to utilize processes related to ceramic, glass, and soft-goods manufacturing, so understanding a variety of processes is important. In addition, compression in development schedules makes the knowledge of rapid prototyping essential. This chapter provides an overview of some of the manufacturing processes along with important terminology.

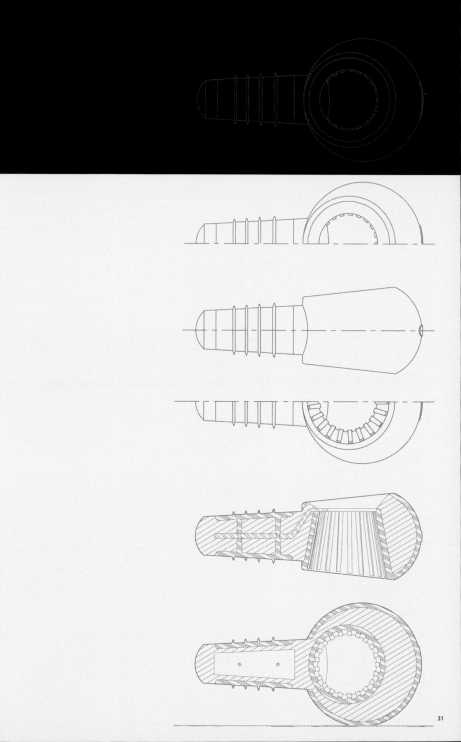

## PLASTICS OVERVIEW

Formed plastics are one of the most common components of a manufactured product. The refinement of the forming process has drastically improved the function and appearance of products, and the technology continues to improve. Each process has unique benefits and restrictions relating to cost, lead time, effect on performance, perception of quality, and durability. Most plastic forming processes utilize either thermoplastic (thermoform) or thermoset plastics (more information is contained in Chapter 4, Material Selection). Thermoplastics soften and melt when heated, while thermoset plastics harden when heated. Designers should work with clients and manufacturers to select the most appropriate material manufacturing process relative to the product objectives. This section includes an overview of common plastic forming, finishing, and joining processes that will help designers make the right choices for the products they are working on.

## TOOLING

The terms tool (or *tooling*) and mold (*molds*) are often interchangeable, and refer to the cavity in which a part is formed. Different forming processes require different types of tooling. Tools that are intended to produce a high volume of parts are made of harder materials (typically steel), while short-run parts may use plastic, aluminum, elastomer, or wooden tools.

The tool is a separate module that fits into an injection molding machine. This image shows an in-progress tool in the tooling shop.

Courtesy of Altitude, Inc.

**Electrical Discharge Machining (EDM) is** a process by which metal tooling is "cut" using electrical discharge. The EDM tool, which is shaped in the form of the desired part, is charged and "pressed" into the blank metal surface. The process slowly removes material in the exact negative of the desired shape. Electrical discharge wire cutting is a variation of the EDM process where a moving charged wire cuts into the metal surface (pictured here).

Courtesy of Altitude, Inc.

**First samples,** also known as first shots or first articles, are initial parts created by the tool and are used for evaluation. Usually, the parts are not textured or in the intended color, and may be missing some **steel safe** features. Steel safe refers to tool details, which can be added later without having to add material to the tool. First samples are used to evaluate sink, mold flow, warpage, shrink rate, and the like.

**Machining** is a process by which material is cut away from a raw piece of material. A specific cutting tool is hand- or computer-controlled for creating a negative impression of a desired surface for forming a part. Many types of materials, including metal, plastic, or wood, may be machined in the creation of a tool. The material selection is dependent on the volume of parts or the level of finish desired. High-volume/high-finish parts require harder tools; low-volume/low-finish parts allow for softer tools.

The machining process usually requires multiple tooling passes, with each pass using a different tool. Early passes "rough-out" the desired geometry, while later passes refine the details.

Courtesy of Scott Models

Courtesy of Altitude, Inc.

This **multicavity** tool is used for housing the remote control for the Malden Mills Polartec Heat Blanket.

**Multicavity molds** are molds that produce multiple parts per cycle. The cavities may be identical, producing many identical parts per cycle (all top housings), or they may be a family of parts intended for the same product (top and bottom housings).

**Single-cavity molds** are molds that produce one part per cycle. Single-cavity tools may be used for low-to-medium volume production, or for very large-scale parts.

Courtesy of Altitude, Inc.

This single cavity mold is used to produce a transformer upper housing for the Malden Mills Polartec Heat Blanket.

**Texture** is important in enhancing the appearance of a product as well as concealing minor surface flaws, such as sink marks, after the tool is cut. There is a wide variety of textures available, ranging from light matte to deep heavy. The desired texture will affect the design of the tooling. The heavier the texture, the more draft is required on the tool. There are many systems of texture specification available, including sources such as Mold-Tech Corporation.

## FORMING (PLASTIC)

**Forming plastic** is the process of transforming a specified material into a desired shape. Raw plastic, or resin, is usually in the form of pellets, powder, liquid components, or sheets. The processes listed below describe the transformation from raw material to manufactured part.

**Blow molding** is a process whereby a softened plastic tube inflates against mold walls and hardens by cooling, typically to form a hollow vessel or container. The three primary types of blow molding are extrusion, injection, and stretch. The selected method is dependent on part size, quantity, and desired level of finish.

**Extrusion blow molding** is a process in which (1) a heat-softened, tubular parison extrudes to the proper length; (2) a two-part mold encloses it, and pinches off the bottom, thereby sealing the parison; (3) air is blown in, expanding the parison against the mold; and (4) after the part cools the mold opens, the hardened part is removed, and excess material is trimmed. Typically, this is the process used for low-volume production on containers greater than 12 ounces.

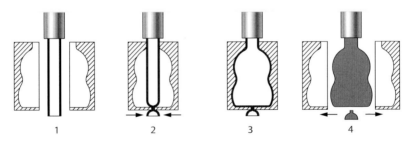

1   2   3   4

EXTRUSION BLOW MOLDING

**Injection blow molding** is a process in which (1) a test tube–shaped injection-molded parison is premolded and transferred into the blow mold while it is still hot; (2) air is blown in, expanding the parison against the mold walls; and the cooled, hardened part is removed from the mold. The process does not require finishing, and yields parts with accurate wall thickness and high-quality neck finish. This is the process used for smaller bottles and containers in high volumes and with high levels of detail and finish.

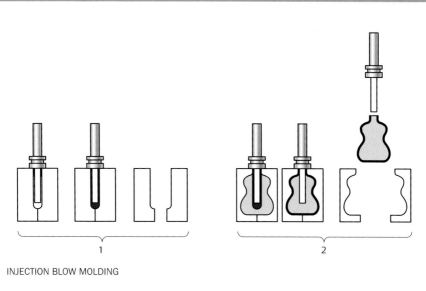

INJECTION BLOW MOLDING

**Stretch blow molding** is a process in which (1) a preformed parison is transferred to a mold while it is still hot, with an internal rod that is part of the mold; (2) as the rod extends into the mold, it stretches the parison, elongating it, and simultaneously blows air into it, expanding it against the mold walls; and (3) the cooled, hardened finished part is removed from the mold and does not require finishing. This is the process used for larger volume containers such as 2-liter beverage bottles.

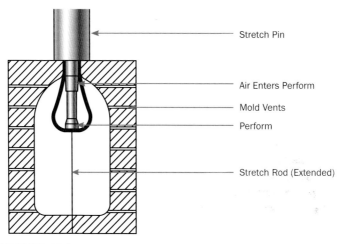

STRETCH BLOW MOLDING

**Casting** is a process by which liquid plastic (usually two-part epoxy) is poured into a mold and allowed to harden. This process is used for low-volume production where highly detailed parts are needed but cost is less important. Tooling may be made of a rigid material or flexible RTV (room-temperature vulcanized rubber). Undercuts may be included in a part when a flexible mold is used, since the flex allows parts to be removed.

**Compression molding** is a process by which (1) a measured amount of thermoset plastic or elastomer is placed in a heated mold and (2) compressed into the desired shape. The mold holds the compressed material until it hardens. Automotive components, container caps (such as cosmetics), and dinnerware commonly use this molding method. Flash may form at the parting line and requires a secondary trim operation. Long dwell times and hand labor usually add to the cost of compression-molded parts.

**Extrusion** is a process by which heat-softened plastic is forced through a die to pro-

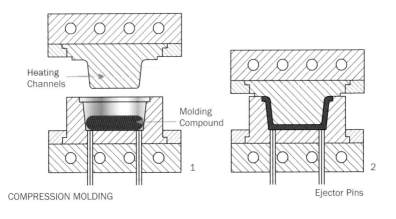

Heating Channels

Molding Compound

1

COMPRESSION MOLDING

Ejector Pins

2

duce a continuous linear part, constant in section. Hard plastic pellets (resin), which are selected based on the desired material and color, move from a hopper through a screw-type driver and are heated to the point of melting, becoming a viscous liquid. The softened plastic is forced through an extrusion die to create the desired section and then is hardened by cooling. Once the extrusion is hardened, it is cut to the desired length.

Uniform wall thicknesses are important to produce a straight extrusion, since varying section thicknesses can result in warpage. Intentional curvilinear extrusions can be achieved by adjusting the die geometry and rate of flow of the plastic. Dual extrusion bonds can also be achieved by feeding different thermoplastic materials through a common extrusion die.

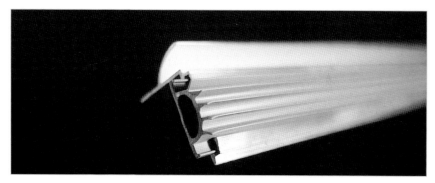

Extruded parts made of different materials can be designed to mechanically interlock. This is an example of an aluminum channel with a mechanically interlocked elastomer part.

**Injection molding** is a process by which melted plastic is injected into a negative cavity and cooled to create a positive part. Injection-molded parts typically have a high level of finish and detail on exterior and interior surfaces.

An injection-molding machine comprises three major functions/components:

**1. Injection** Hard plastic pellets (resin) move from a hopper through a screw-type driver and are heated to the point of melting and becoming a liquid. The liquid plastic is forced through a nozzle into the mold. The pellets are selected based on the desired material and color.

**2. Mold** A specially designed metal tool consisting of a "mold base" and one or more cavities. The cavities are contoured to the exact shape of the designed final part. The liquid plastic is forced into the cavity void where it cools and hardens. The mold can vary widely in size and complexity. The mold is removable, allowing one injection-molding machine to use different molds to produce different parts.

**3. Clamping** applies force to open and close the mold closed during the injection process. When the new plastic part has cooled and hardened, the mold is opened and an ejector forces the part from the mold.

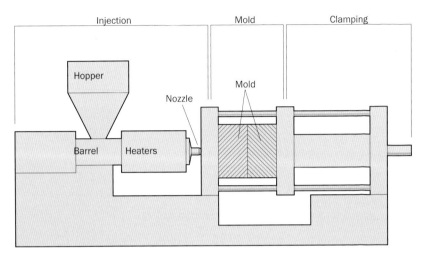

Plastic resin in the form of solid pellets is contained in the hopper. The pellets move through the barrel, are heated until they become liquid, and are injected into the mold, where the resin cools and hardens, forming a solid part. The mold is opened and the part is ejected.

When designing an injection-molded part, designers should ensure that there is adequate draft (angle in the part to allow easy removal from the mold), there are no undercuts (which would make it impossible to remove from the mold), sharp corners are avoided (which may not fill properly), and a uniform wall thickness is maintained (to prevent uneven cooling, which can result in warpage). It is possible to include undercut-like features using shut-offs, slide-actions in the mold, cam blocks, or collapsible cores. These features add cost and complexity to the mold, but can add value, reduce parts count, or reduce labor costs. Variable wall thicknesses are possible, provided the thickest section is less than fifty percent thicker than the nominal (typical) wall section.

Final plastic parts will have a visible "gate," which is the point where the liquid plastic was injected into the mold, and surface marks were created by the ejector pins. These features are designed to be on a less apparent surface (like the inside of the housing). Knit-lines (where two or more flows of melted plastic meet, cool, and harden) may also be visible if the pellets are not thoroughly mixed or if resins with metallic particles are used. Many parts also include structural ribs or raised bosses that assist in aligning with other parts and mounting components.

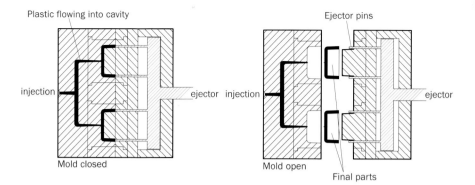

One injection-molding machine can produce many different types of parts by "changing out" the mold. The mold is designed to produce a specific part and interface seamlessly with the machine. Molds may have multiple cavities and vary in complexity.

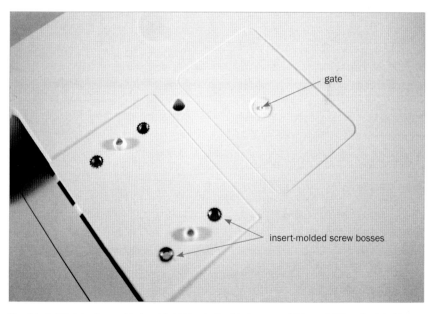

The gate in this injection-molded part is hidden in the label recess of this part. When the label is applied, the gate will not be visible. This part also contains threaded metal inserts.

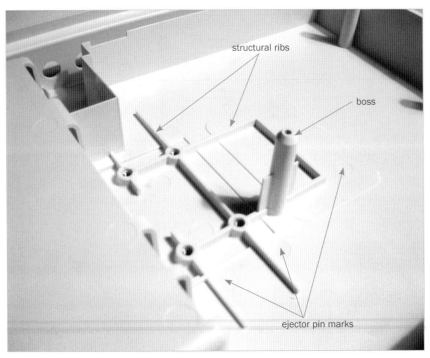

structural ribs

boss

ejector pin marks

Ribs and bosses add structure and provide component-mounting points. Marks left by the ejector pins are hidden on interior surfaces.

Injection molding can also accommodate many specialized processes. For example, it is possible to insert a specially designed part into the mold prior to injecting the liquid plastic (insert molding). The plastic hardens around the inserted part, resulting in a final part that is a combination of two materials, such as a housing with metal screw bosses. Also, because of the elasticity of plastic, built-in closure systems (living hinges and snap closures) can be designed into a part. It is also possible to mold a part in hard plastic, transfer it to a separate mold, and inject elastomer materials (overmolding) over entire surfaces or in specific locations.

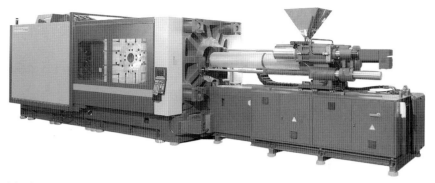

Injection-molding machines come in many shapes and sizes depending on the size of part produced, the type of material, and the cycle time, or the rate at which parts are produced.

**RIM (Reaction Injection Molding)** is a process in which liquid resin and a catalyst are mixed in an injection-molding cavity. The two components chemically react to form a rigid part. This is the process used for automobile bumpers, electrical enclosures, enclosures for medical equipment, and other large, complex low-volume parts. This process allows for easy inclusion of inserts, is extremely rigid, and is wear resistant.

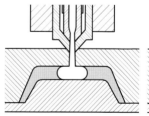

First liquid component is injected into mold.

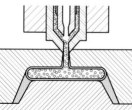

Second liquid component is injected into mold.

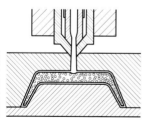

Final part has smooth surface and foam core.

**Rotational molding** is a process by which a hollow mold is loaded with plastic resin, liquid, pellets, or powder, then heated externally and rotated to distribute the material evenly on the internal surface of the mold. When the mold cools, the rotation is stopped and the hardened part is removed. Parts produced with rotational molding have a highly finished exterior surface but limited sharpness in detail. The process is ideal for larger parts with no internal part features.

After a form is created that satisfies the users' functional needs, simple considerations can help yield a higher-quality part. When designing a part for rotational molding, a designer should avoid large, flat surfaces, which have a tendency to warp. By using features such as ribs, crowns, steps, or domes, warpage can be minimized. Textured surfaces are also helpful in masking surface warpage (a glossy surface exaggerates slight surface variations). It is also recommended that the spacing between two parallel walls be greater than five times the nominal wall thickness of the part. Rotational molding tooling costs are relatively low, but cycle times are long (the process yields fewer parts in a given time than other processes).

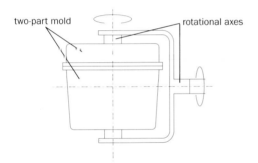

two-part mold          rotational axes

The Rotational Molding Process 1) The two-part mold is loaded with a premeasured amount of resin in liquid, powder, or pellet form; 2) the mold assembly is heated in a large oven, while rotating on two axes, allowing the material to evenly coat the inside surface of the mold; 3) While continuing to rotate, the mold assembly is removed from the oven. The mold is then opened and the solid part is removed, and if necessary secondary operations are performed.

**Thermoforming** is a process by which a sheet of plastic is heated and forced into a mold cavity using vacuum or pressure (depending on the specific process).

When designing a thermoformed part, there are important considerations. First, the depth of the draw should be no greater than the largest dimension across the face. Next, an inside corner radius should be no less than 25 percent of the wall thickness. Finally, it is preferred that parts have a draft angle of one degree or greater. Thermoformed parts typically require secondary processes for trimming the part from the formed sheet or adding holes or slots. These processes are usually automated. Tooling costs are relatively low as molds can form from less expensive and more easily machined materials such as wood or aluminum.

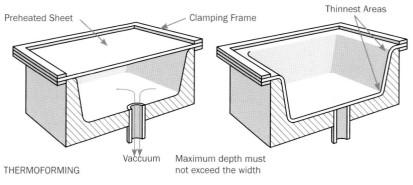

Preheated Sheet    Clamping Frame    Thinnest Areas

Vaccuum    Maximum depth must not exceed the width

THERMOFORMING

The thermoforming process comprises three stages: 1. Sheet plastic is clamped securely in a frame and heated (softening the plastic) then lowered to contact the mold. 2. Vacuum or pressure forces the heated plastic against the mold surface. 3. The cooled, formed sheet is withdrawn from the mold and released from the frame.

**Transfer molding** is a process in which thermoset plastic or elastomer is placed in a transfer chamber, heated, then injected into a mold cavity, where it cools and hardens. Because inserts are easy to use with this process, it is commonly used for two component parts, such as a rubber bushing molded around a fastener.

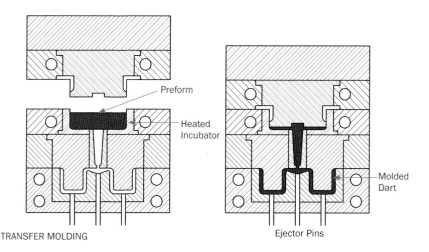

Preform

Heated Incubator

Molded Dart

TRANSFER MOLDING    Ejector Pins

**Urethane castings** are produced using silicone or RTV (Room Temperature Vulcanization) molds. In some cases, a master pattern is created, scaled to compensate for shrinkage. In most cases, draft is not an issue and small undercuts are not a problem because the silicone mold is flexible. Cast urethane parts closely simulate final injection-molded plastic parts. The quality and finish of the final parts are identical to the level of finish of the master part; urethane castings can be molded in a selected color or painted.

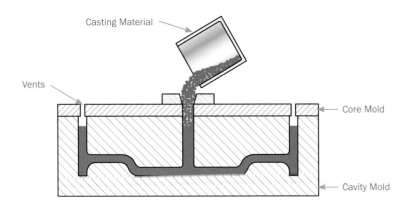

Casting Material

Vents

Core Mold

Cavity Mold

## FINISHING (PLASTIC)

After a part has been formed, it may be subjected to additional processes for improving performance or appearance.

**Electroplating** is a process by which a plastic part is coated with a layer of copper, nickel, or chrome using electrical deposition. The result is a glossy, durable, bright metallic finish. Because plastic does not conduct electricity, parts must be precoated with a conductive material.

The conical reflector of a flashlight is a good example of the type
of finish offered by the electroplating process.

**Flocking** is a process by which short, fine fibers are applied to an adhesive coated
surface for achieving a soft, fuzzy feel. The fibers are applied electrostatically or
mechanically.

**In-mold decorating** is a process of applying graphics to a formed plastic part. First, a clear sheet of plastic is printed and thermoformed to match the contour of the final part. This part is inserted into the mold prior to the injection of liquid plastic (in the case of injection molding), which bonds the sheet to the part. The result is a printed image sandwiched between a clear layer of plastic and the primary plastic part, creating an extremely wear-resistant graphic image.

**Steps in the IMD Process**

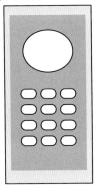

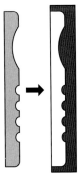

1  Film is screen printed (decorated) on first and/or second surface.

2. Film is formed to shape and trimmed to final size.

3. Film is inserted into special mold tooling and part is injection molded, permanently bonding decorated film.

**Labels** Information such as logos, instructions, or warnings can be printed on paper, plastic, or metal and attached to the surface of a formed part. Labels may be multi-colored and preprepared with adhesives. In addition, formed parts may include a recessed area matching the depth and shape of the label. Typically, this "pocket" is not textured in order to increase label adhesion.

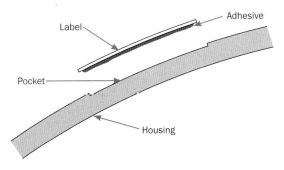

**Pad printing** is a process for printing graphics on a complex surface. Pad printing involves transferring ink from a printing plate etched with the desired image to the part by way of a silicone pad. Because the silicone pad is flexible, it can conform to the contour of the part printed without distorting the image. Extremely fine detail and tight multicolor registration are achievable. Almost any material can be pad printed, so long as compatible ink is used.

**Painting** If a desired color or finish is not achievable with existing plastic resin, a part may be painted to enhance its appearance or function. One situation where a designer might choose painting is in the case of a plastic part with a metallic finish. Metallic finish resins are available. However, knit lines are often visible in these parts. A painted metallic finish would offer an even finish across the entire surface. Also, soft-touch paints are available which provide a nonslip surface at a lower cost than overmolding.

**Screen printing** is a process of applying graphics to a part using a silkscreen. The process is high quality and inexpensive; however, registering the color screens and allowing drying time typically limits the number of colors to two.

**Sputtering** is the process of applying a metallic coating to a substrate by causing atoms to break away from a target metal and form a vapor that is then deposited on the plastic part surface in a thin, even coating. There is no melting of the metallic source during this process. Sputtered parts are durable, lightweight, and inexpensive to produce.

**Vacuum metallization** is the process of heating a finishing metal to the point of vaporization, then allowing it to deposit on a plastic substrate. Only pure metals can be used in this process, which results in a highly polished surface. Vacuum metallization is used for interior applications because the finished layer is prone to corrosion.

**Water transfer printing** is a process of adding a printed film coating to a formed substrate. An image is printed on a water-soluble film that is placed on the water's surface. The film dissolves, leaving the printed image floating on the water's surface. The substrate is dipped into the water, causing it to be coated by the printed image. After it is cleaned and dried, it is coated with a clear finish. This process is commonly used for products that have detailed images covering the entire surface, such as decorative cell phone face plates or imitation wood grain in an automobile.

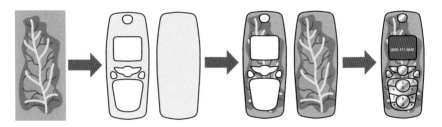

WATER TRANSFER PRINTING

## JOINING (PLASTIC)

The forming of plastic parts is usually followed by joining multiple parts together to create assemblies with certain functionality, or to contain functional components. The way parts are joined can vary widely based on functionality, forming processes, use conditions, reparability, durability, and cost. Assembly is usually one of the most costly aspects of producing a product. Therefore, carefully choosing the methods of joining parts during the design process is critically important. The joining of parts is determined well in advance of creating tooling, so that the appropriate features to aid in specific joining methods can be designed into the parts. This section describes common methods for joining plastic parts.

**Adhesive bonding** is the process of joining two or more parts by using an adhesive. Often when adhesive bonding is used, the parts include mechanical features that prevent misalignment. Prior to bonding, parts are usually cleaned to ensure that mold releases are removed to ensure a secure bond. Adhesive bonding is compatible with most plastic-forming processes.

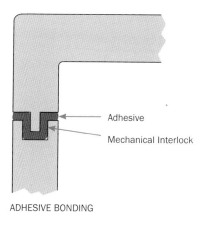

Adhesive

Mechanical Interlock

ADHESIVE BONDING

**Fasteners,** such as screws, are often used to attach components to housings and for joining multiple housings. This method of joining allows parts to be disassembled for repair or recycling. Injection-molded parts may have molded-in features (bosses) that accept the screws. Self-tapping screws are often chosen when assembling plastic parts because they tap into the plastic, creating a matching thread pattern. Thermoformed parts may require inserts to accept the screws.

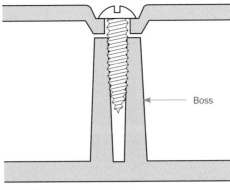

Boss

FASTENER

**Snaps** are molded-in features that allow parts to join by interlocking. Snaps are designed into a molded part and take advantage of the flexibility of plastic. This method allows for the quick assembly of parts at a very low cost. This type of fastener is used for joining injection-molded parts.

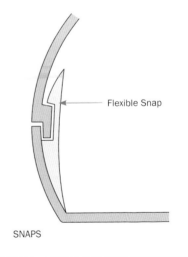

Flexible Snap

SNAPS

**Ultrasonic welding** is the process of joining parts using ultrasonic vibration to create friction between two parts, causing them to heat up, melt, and fuse together. The edge of one of the mating molded plastic parts contains a thin ridge of plastic (energy director), which focuses and controls the area of the weld. This method of joining yields a strong bond, enhancing the rigidity of a housing assembly. Once parts are welded, they cannot be disassembled. This is the process used for permanently joining injection-molded parts.

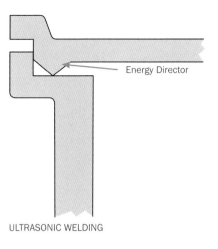

ULTRASONIC WELDING

## METALS OVERVIEW

The use of metal in manufactured products predates the use of plastics. In many cases, plastic components have replaced metal ones to make products lighter and less expensive. However, metal components remain critical to the aesthetics and functionality of many products. Metal may be used for its structural or conductive properties or purely to add perceived value through its look and feel. Different processes are used to form these components, and each process has unique benefits and restrictions relating to cost, lead time, effect on performance, perception of quality, and durability. Most metal forming processes utilize either ferrous or non-ferrous metals (more information is contained in Chapter 4, Material Selection). Ferrous metals contain iron and have a high melting point, while nonferrous metals tend to have lower melting points and are more conductive. Designers should work with clients and manufacturers to select the most appropriate manufacturing process and material relative to the product objectives. Included in this section is an overview of common metal forming, finishing, and joining processes.

## FORMING (METAL)

Forming is the process of transforming a specified material into a desired shape. This section describes processes for forming cold and molten metal.

**Casting** is a process by which molten metal is poured into a mold and allowed to harden.

**Sand casting** involves creating a master pattern that is pressed into a bed of sand mixed with a heat resistant binder, resulting in a single-use (expendable) sand mold. The mold includes a sprue, which channels molten metal into the part cavity, runners for distributing the metal throughout the mold, and vents that allow gases to escape. After the metal cools and hardens, the mold is broken away to remove the part. Sand casting requires secondary processes to trim away the sprue, runners, and other extraneous parts.

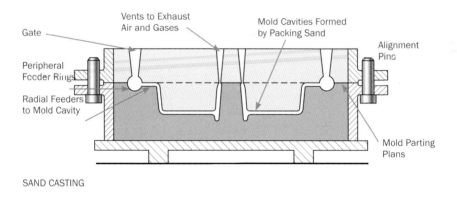

SAND CASTING

**Investment casting** involves creating a master pattern made of a material that melts or evaporates when subjected to heat. The pattern is packed in sand or dipped in ceramic slurry. In the situations where ceramic is used, it is allowed to harden and the mold is inverted, then heated, allowing the pattern to melt away. Molten metal is poured into the mold. After the metal cools and hardens, the mold is broken away. Investment cast parts require secondary processes to trim away extraneous parts.

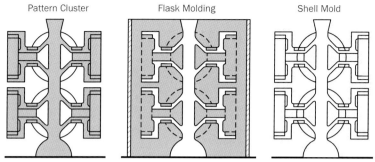

Pattern Cluster          Flask Molding          Shell Mold

INVESTMENT CASTING

**Die casting** involves forcing molten metals into reusable molds using pressure. Metals with lower melting points (aluminum, copper, zinc, etc.) are used with this process. When the parts cool and harden, they require secondary operations to remove flash.

**Contour roll forming** is a process by which metal strips or sheets are converted into continuous section profiles through a sequence of mating rollers. Each set of rollers in the sequence progressively bends the strip until it reaches the desired section. This process adds strength and rigidity to flat materials.

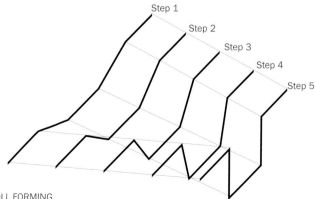

Step 1
Step 2
Step 3
Step 4
Step 5

CONTOUR ROLL FORMING

**Deep draw (stamping)** is a process that produces hollow shapes from flat sheets of metal. The process involves a punch pushing a blank sheet of metal into a cavity. Secondary trim operations are often necessary.

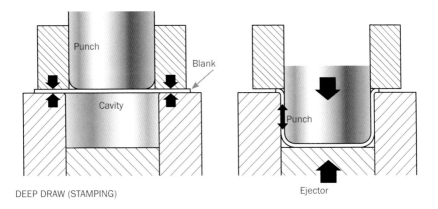

DEEP DRAW (STAMPING)

**Extrusion** is a process by which molten metal (usually aluminum) is forced through a die to form a long continuous section profile. Tooling costs are lower when compared to other processes, and little scrap is produced. Designing a part with a continuous section is important in preventing curvature in length. Extrusions can be cut to a desired length, and secondary operations are used to polish or add features through material removal.

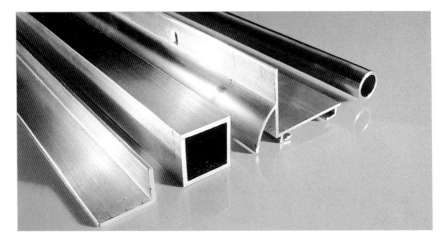

**Forging** is a process of compressively deforming metal (heated or cold) between two dies using impact or pressure. Forging can be used to form parts of almost any size and shape and is commonly used for tools and fasteners.

**Hydroforming** is a process for creating complex formed metal parts, without the need for costly matched die pairs. The process utilizes pressurized liquid to force sheet metal into a single die cavity.

**Spinning** is a process for producing hollow rounded shapes by forcing a spinning metal blank against a pattern. This process yields accurate and cost-effective parts.

SPINNING

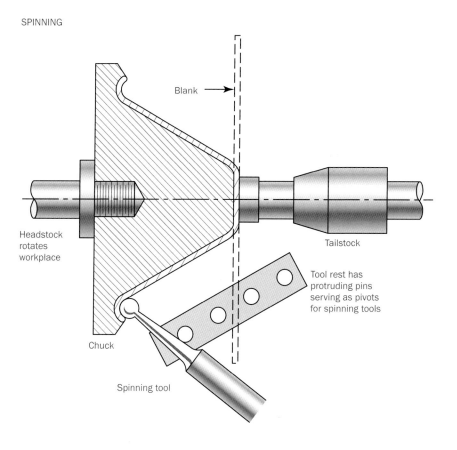

Blank

Headstock
rotates
workplace

Tailstock

Tool rest has
protruding pins
serving as pivots
for spinning tools

Chuck

Spinning tool

## FINISHING (METAL)

After a part has been formed, it may be subjected to additional processes for improving its appearance, performance, or preventing corrosion.

**Abrasive finishing** is a process for finishing the surface of metal parts for removing small amounts of material from the surface to improve the function or appearance. These processes include polishing, buffing, wire brushing, and burnishing.

**Anodizing** is an electrochemical process that thickens and toughens the naturally occurring protective oxide on metal parts. The anodic coating becomes part of the metal and is porous enough to allow adherence of secondary coatings for coloring and protection. This process can be used on a variety of metals, with aluminum most commonly anodized.

**Coating** includes a number of processes for covering the surface of the metal to improve performance, prevent corrosion, or improve appearance.

**Electrostatic painting** involves positively charging paint particles and negatively charging a metal part. The particles are attracted to and then adhere to the part, resulting in a very efficient use of paint and even coating.

**Metallic deposition** is a process for applying thin layers of metal to the surface of a part. These processes include electroplating, sputtering, and vacuum metallizing. Each process has unique benefits and restrictions.

**Porcelain enameling** is a process by which enamel is applied to the surface of a metal part then fired to harden the coating and fuse it to the surface. Porcelain enamel coatings are extremely durable and fade resistant.

**Powder coating** is a process by which parts are coated with a fine thermoplastic powder then heated until the powder melts. This process causes the particle to fuse to the surface of the metal, providing a durable coating. This process often uses electrostatic charging of the powder and part to evenly coat and efficiently use the powder.

**Labels** with logos, instructions, or warnings can be printed on paper, plastic film, or metal plates and attached to the surface of a formed part, usually by using adhesives.

**Pad printing** is a process for printing graphics on a complex surface. Pad printing involves transferring ink from a printing plate, etched with the desired image, to the part by way of a silicone pad. Because the silicone pad is flexible, it can conform to the contour of the part printed without distorting the image.

Extremely fine detail and tight multicolor registration are achievable. Almost any material can be pad printed, so long as compatible ink is used.

**Screen printing** is a process of applying graphics to a part using a silkscreen. The process creates high quality and inexpensive graphics. The difficulty of registering the color screens and allowing drying time typically limits the number of colors to two.

## JOINING (METAL)

Metal forming is often followed by joining several parts together to create certain functionality or contain functional components. The way they are joined can vary widely, based on the function, forming process, use conditions, reparability, durability, and cost. Assembly is usually one of the most costly aspects of producing a product. Therefore, carefully considering the joining method during the design process is critically important. The joining methods are determined well in advance of creating tooling so that the appropriate features to aid in specific joining methods can be designed into the parts. This section describes common methods for joining metal components.

**Adhesive bonding** is the process of joining two or more parts by using an adhesive. Often when adhesive bonding is used, the parts include mechanical features that prevent misalignment. Prior to bonding, parts are cleaned to ensure that coatings are removed to ensure a more secure bond.

**Brazing** is a method of adhering metal components with metallic adhesives. The adhesive, a brazing alloy, is formulated to melt well below the temperature of the metals being brazed to avoid thermal distortion. When the brazing alloy is heated, it flows into joints with the aid of flux, which also prevents oxidation and aids in bonding. Brazing usually requires temperatures over 800°F (427°C).

**Mechanical** joining may be permanent or removable depending on functional needs. For frequently replaced parts, or for access to areas that need regular maintenance or repair, removable fasteners are advantageous. When weight and assembly costs are more important, permanent fasteners may be more desirable.

**Collar** This device permits free rotation of a shaft while preventing it from moving axially. A collar is usually a metal ring with an integral setscrew. The ring slips snugly over the shaft, and the setscrew is used to tighten the collar by placing the shaft in compression and the ring in tension to one another.

Setscrew

**Integral joiners** include features designed into metal parts that allow them to mechanically interlock. Actions such as lancing, notching, and folding can result in features that mechanically interlock two metal parts with no additional fasteners.

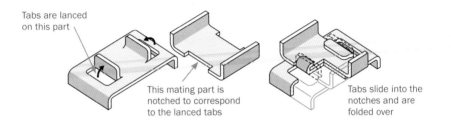

Tabs are lanced on this part

This mating part is notched to correspond to the lanced tabs

Tabs slide into the notches and are folded over

**Pins** are used to permit free rotation of a shaft while preventing it from moving axially. The geometry of a pin can vary, but they are typically long and cylindrical. Pins are inserted into holes that are drilled perpendicular to the axis of a shaft.

Cotter pin

Shaft

Collar

**Retaining rings** are used to permit free rotation of a shaft while preventing it from moving axially. A retaining ring is usually a flat metal C-shaped part that is snapped into a groove cut around the diameter of a shaft.

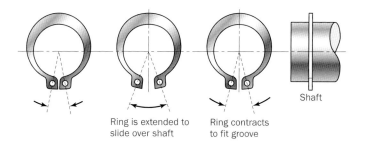

Ring is extended to slide over shaft

Ring contracts to fit groove

Shaft

**Rivet** There are many types of rivets, most of which are designed to permanently connect components. Many types allow blind assembly (when the components can only be accessed from one side).

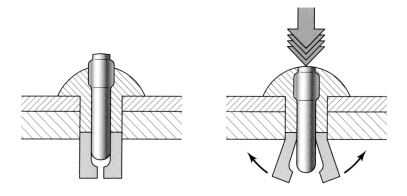

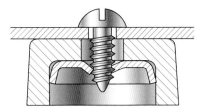

**Screw** These may include many types of fasteners, including self-tapping screws, machine screws, and bolts. Many screws are used in conjunction with nuts, which help keep the screws securely in place.

**Staple** A U-shaped fastener formed from high tensile strength steel wire. Staples pierce, in two places, the materials that they are intended to fasten.

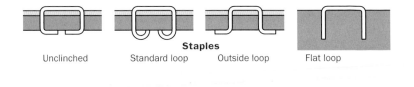

**Staples**

Unclinched          Standard loop     Outside loop          Flat loop

**Snap** A two-part fastener comprised of lanced tabs on one component and corresponding holes on the second component. When the two parts interface, the tabs deflect until they align with the holes, where they snap back to their original form and secure the assembly.

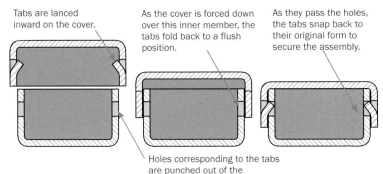

Tabs are lanced inward on the cover.

As the cover is forced down over this inner member, the tabs fold back to a flush position.

As they pass the holes, the tabs snap back to their original form to secure the assembly.

Holes corresponding to the tabs are punched out of the sidewalls of the inner member.

**Soldering** A process by which a nonferrous, low melting point filler (solder) is used to join metal components. When the solder is heated, it flows into joints with the aid of flux, which allows the solder to flow, prevents oxidation, and aids in bonding. Soldering usually requires temperatures under 800°F (427°C).

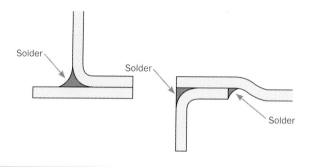

---

**Welding** The process of joining metal parts by applying heat, at times pressure, and/or a filler metal having a high melting point. The process results in a joint as strong as or stronger than the metal parts themselves. Specific types of welding are chosen based on part geometry, thickness, material, cost, and access. The two basic categories of welding are fusion and solid state welding. Within each category variations of the process exist.

---

**Fusion welding** A process in which the temperature of the metal to be welded is brought to its melting point and joined, with or without filler metal. Processes include arc, electron, laser, and resistance welding.

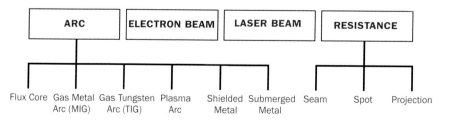

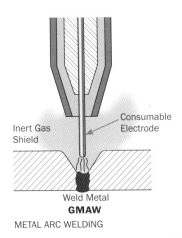

Gas metal arc welding uses a wire electrode, which is fed through a manually manipulated gun or torch. The consumable electrode acts as filler material, and the weld area is surrounded by an inert gas used to prevent oxidation.

Inert Gas Shield

Consumable Electrode

Weld Metal

**GMAW**

METAL ARC WELDING

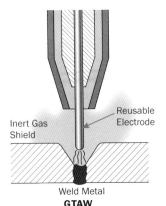

Gas tungsten arc welding uses a tungsten nonconsumable electrode to heat and weld together the work pieces. An inert gas surrounding the weld prevents oxidation, and, if needed, a consumable filler rod is added manually to provide molten metal to the weld.

Inert Gas Shield

Reusable Electrode

Weld Metal

**GTAW**

TUNSTEN ARC WELDING

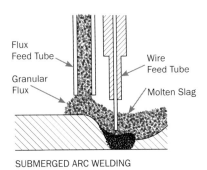

Submerged arc welding uses granular flux to cover a puddle of molten metal deposited by a gas metal arc welder electrode. The flux serves as a barrier that prevents heat from escaping, enabling thicker materials to be welded. This is typically a semi- or fully automated process.

Flux Feed Tube

Wire Feed Tube

Granular Flux

Molten Slag

SUBMERGED ARC WELDING

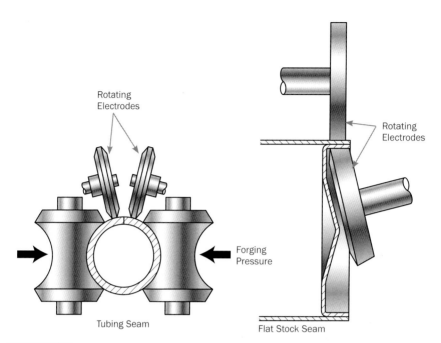

Rotating Electrodes

Rotating Electrodes

Forging Pressure

Tubing Seam

Flat Stock Seam

## SEAM WELDING

In the seam welding process, two parts are welded using opposing rotating electrode wheels, which create a weld along continuous abutted edges.

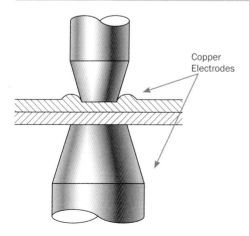

Copper Electrodes

In the spot welding process, overlapping sheets of metal are squeezed between two copper electrodes that weld the metal together in that localized area using an electrical current. Spaced in measured intervals, spot welds appear as slight deformations on the metal surface.

## SPOT WELDING

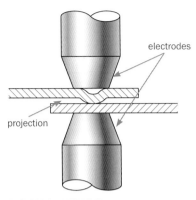

In the projection welding process, projections, or protrusions, are formed, forged, or machined into a metal part. These features concentrate heat, caused by electrical flow, to melt and join parts.

PROJECTION WELDING

**Solid state** is a group of welding processes that join metals at temperatures below the melting point of the base materials being joined, without the addition of filler metal. Pressure may or may not be used, and there is very little melting of the base metals. This process is useful when dissimilar metals are joined and thermal expansion and conductivity are less important. One example of solid state welding is friction welding (see diagram at right). This process does not use an external source of heat; rather it relies on heat generated by the friction of two parts rubbing together. In this example, two rods or pipes are joined by heat generated from the friction of one part rotating against another.

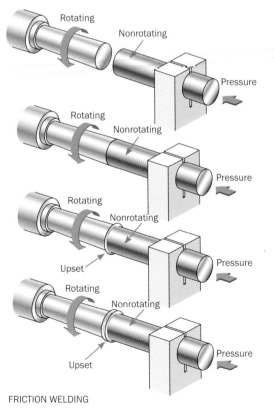

FRICTION WELDING

## CERAMIC OVERVIEW

Ceramic is a continuously evolving material that is commonly used in the design and manufacturing of consumer, commercial, and architectural products. Contemporary manufacturing of ceramics is focused on refining traditional processes and developing new processes and clay formulations for high-tech uses.

After a design is developed, an appropriate **clay body** (see materials selection) and forming process is selected. These choices are made based on the geometry of the part and the desired properties. This section provides important terminology and definitions related to ceramic manufacturing.

**Bisque** Clay that has been fired to red hot but has not yet been glazed.

**Bisque firing** The first firing of clay objects without glaze (1560°F–1830°F [850°C–1000°C]). The rapid rise in temperature at the beginning of this firing will cause steam pressure to build up inside the clay, and sometimes shatters it. In this firing, the bound water that is part of the structure of clay is driven off, changing the clay into a ceramic material that is unaffected by water.

**Bone dry** State of dryness of greenware when it is ready to be bisque fired.

**Clay body** Generally refers to a combination of clay ingredients calculated to mature at a desired temperature and to have desired working or color characteristics.

**Dry pressing** Dry powder clay formed with pressure in a mold. The process eliminates drying time prior to firing, enabling quick turnaround, no shrinkage, and little warping. Typically used with plates and other low-profile vessels.

The pressure may be applied **uniaxially** (in one axis) or **isostatically** (from all directions). Uniaxial forming is the most common dry press process and is used to form tiles or plates. Isostatic forming involves applying pressure with a shaped membrane or "bag." This process results in more uniform density and greater compaction and is used for high-volume dinner plates or cups.

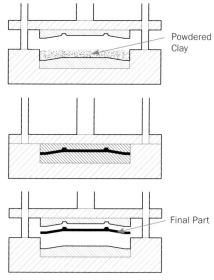

Powdered Clay

Final Part

DRY PRESSING

**Extrusion** The plastic clay is forced through a die to create a continuous length part with a consistent cross section. The extruded profile is cut to the desired lengths. This is the process used for producing bricks and pipes.

**Firing** The process of converting clay to ceramic. It involves heat of at least 1100°F (600°C). During firing, clay is changed into a stone-like material as organic matter is burned away. Also, during firing, the colors of the minerals used in clays and glazes usually change, almost always in a predictable way.

**Glaze** Finely ground minerals suspended in a liquid, which is applied by brushing, pouring, or spraying on the surface of bisque-fired ceramic ware. After the glaze dries, the ware is fired to the temperature at which the glaze ingredients will melt together to form a glassy coating that fuses to the surface.

**Glaze firing** The second firing, to the appropriate temperature at which the glaze materials will melt to form a glasslike surface coating on the ceramic material. This firing is usually at a considerably higher temperature than the first bisque firing.

**Greenware** Unfired ceramic objects.

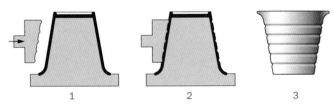

1          2          3

JIGGERING SEQUENCE

**Jiggering** A process that uses wet clay prepared by extrusion, which is then placed on a mold and formed against a rotating mold. The flat disk is placed on a plaster mold in the shape of the piece to be formed. The mold may contain decorative surface relief. After forming, the piece is dried on the mold, removed, and fired. The process requires a large number of molds for production (since the parts dry on the mold).

**Kiln** A furnace for firing ceramic objects. Kilns are made of refractory and insulating materials and are heated by electric elements or a variety of combustible fuels.

**Leather-hard** Dried clay that is stiff but pliable and damp enough to be joined to other pieces of clay with bonding slip. This is the ideal state for pots to be trimmed. At this stage of drying, the shrinkage of the clay is largely complete.

**Overglazing decoration** A ceramic or metallic glaze decoration applied and fired on a previously glazed surface of ceramic ware. Overglaze firings are at a lower temperature than the original glaze firing.

**Porosity** 1. The ability of a fired body to absorb water by capillary action. 2. A measure of the proportion of pores in a ceramic material.

**Press molding** Forming objects by pressing plastic ceramic body onto and into absorbent plaster molds. As the clay dries and hardens, it shrinks slightly and can be removed from the mold.

**Pressure casting** A process similar to slip casting, but with pressure applied to the slip in the mold. This helps force water from the slip through the part and into the mold where it is absorbed. Increasing the slip pressure dramatically cuts the time required to achieve the desired wall thickness.

**RAM pressing** A process in which a billet of clay is placed on the lower mold and is pressed to shape by the upper mold. This is the process used to form large platters, particularly oval ones that cannot be jiggered easily, and large floor tiles.

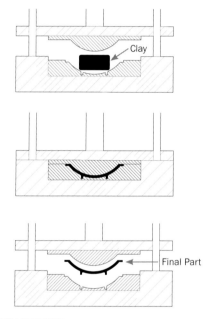

RAM PRESSING

**Shrinkage** Contraction of the clay in either drying or firing. Dry shrinkage is reversible with the return of hydration, but shrinkage due to chemical and physical changes in the clay caused by firing is permanent.

**Slip** Clay in liquid suspension.

**Slip casting** A process in which (1) a plaster mold is created that represents the intended external geometry of the part; (2) next, clay slurry is poured into the plaster mold. (3) As water from the slurry is absorbed into the mold, the clay begins to collect on the inside mold surface. (4) Once the desired wall thickness is achieved, excess clay slurry is poured out. (5) The clay part is removed from the mold, dried, and fired. This process is economical and allows for easy duplication of complex forms and decoration using the same mold.

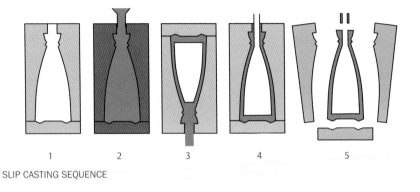

SLIP CASTING SEQUENCE

**Vitreous** Pertaining to the hard, glassy, and nonabsorbent quality of a body or glaze.

**Warping** Deformation of a clay shape caused by uneven stresses during shaping, drying, or firing.

## GLASS PROCESSES

Glass processes are used in the design and manufacturing of consumer, commercial, and architectural products. Contemporary manufacturing of glass is focused on refining traditional processes and glass formulations for high-tech uses.

After a design is developed, an appropriate glass type and forming process is selected. These choices are made based on the application and desired properties. This section provides important terminology related to glass manufacturing.

**Automated glassware forming** The production of glassware items, such as wine glasses or tumblers, is an automated variation of traditional glass blowing. A pipe is used to gather molten glass, form a parison, and blow the glass into a mold. The forming process is followed by finishing processes, which include stem forming or waste glass trimming.

**Blow and blow** A process in which molten glass is placed into a mold where a parison with a finished neck is created. Air is used to form a parison, which is removed from the mold and transferred to a blow mold, where air is used to form the final shape.

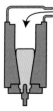

| 1. | 2. | 3. | 4. | 5. | 6. |
|---|---|---|---|---|---|
| Molten glass dropped into blank mold | Neck formed | Blank blown | Blank shaped | Blank transferred to blow mold | Final shape blown |

BLOW AND BLOW

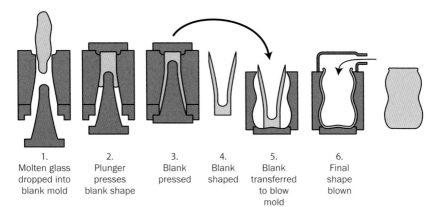

| 1. | 2. | 3. | 4. | 5. | 6. |
|---|---|---|---|---|---|
| Molten glass dropped into blank mold | Plunger presses blank shape | Blank pressed | Blank shaped | Blank transferred to blow mold | Final shape blown |

PRESS AND BLOW

**Container forming** Automated container manufacturing is a multistage process. First, raw glass is heated until molten. Next, the glass is formed in two steps. Then, containers are annealed and cooled. Finally, the containers are inspected and labeled. There are two common forming processes for glass containers.

**Glass blowing** Historically, glass blowing has been the most common method of forming glass. A hollow pipe is dipped into molten glass. The pipe is rotated to gather glass and then rolled on an iron slab to form a parison. Through the process of rolling on the slab, reheating, and blowing air into the tube the parison is shaped into a hollow form. Iron or wooden molds are used to create the final shape of the object.

**Press and blow** A process in which molten glass is placed into a mold and is pressed by a plunger to create a parison. The parison is removed and transferred to a blow mold, where air forces the glass into the final shape.

**Secondary processing** Part of the production process in which some types of glass are subjected to secondary processing such as **annealing**, which involves controlled cooling processes intended to prevent cracking; **tempering**, which involves uniform reheating and rapid cooling to induce tension in the glass; and **coating**, which involves physical and chemical processes that change optical qualities, enable scratch resistance, or increase strength. **Decorating** processes involve material removal (physically or chemically), adding material (low temperature glass, metal, or enamel coatings, films, or printing), and **postforming**, which involves selective reheating and manipulation of forms and surfaces.

## RAPID PROTOTYPING

Overview Selecting a method of prototyping a three-dimensional product is dependent on multiple factors, including purpose, cost, and time available to produce it. In choosing a process, consider if it is for an appearance model, an ergonomic study, or a mock-up of a mechanism. For many applications, a fast, inexpensive prototype with low finish level works very well, but in some cases, a realistic representation is critical.

**3D Printing** On a vertically articulating platen, a layer of wax, ink, or glue is printed on a thin layer of build material such as starch or plaster, which then solidifies and hardens. The configuration of each printed layer is determined by a 3D CAD data file. The platen indexes down and another layer of build material is laid across the previous one, resulting in a thicker combination. This process is repeated many times until the entire three-dimensional part is built. The print material is usually available in different colors, and in some devices, multiple colors can be printed in a single part. Some materials require a secondary process, such as impregnating the part with wax, glue, or resin. Although these processes are relatively inexpensive and fast, the finish and resolution can be low.

A Thermo-Jet wax-based 3D printer

Courtesy of Altitude, Inc.

**Computerized Numerical Control (CNC)** A process in which parts are produced using a CNC machine tool to cut the part from metal, plastic, or composite stock. Starting with 3D CAD data, a milling sequence is developed that performs the actual machining. This sequence may use one tool for the entire milling, or may require tool changes to achieve finer details and surface quality. Depending on the shape of the object to be machined, one or more fixtures may be required.

The primary advantage of CNC prototypes is that they can be produced in the exact material planned for the final product. This process results in the creation of fully functional prototypes that can be used for testing in the planned product environment. The primary disadvantage of CNC prototypes is that the time required and the cost of the prototype is highly dependent on the size and geometric complexity of the piece.

Courtesy of Scott Models

**Fused Deposit Modeling (FDM)** This process is embodied in a small desktop machine that may run in an ordinary office environment. Based on a 3D CAD data file, the machine builds up thin layers of liquid resin one after another. As each layer hardens, the next is added. When the build is complete, the prototype is removed from the build platform and the support material is removed. FDM systems often allow different material and color choices. FDM prototypes are not suited for highly finished parts or as master patterns because of the poor quality and resolution produced.

**Laminated Object Manufacturing (LOM)** A process that involves laminating sheets of paper or plastic that have been laser-cut to a specific shape based on a 3D CAD data file. The sheet material in roll form advances to the cutting platform, and a laser beam cuts one layer of the desired shape. The material advances, and an uncut layer is heated or pressure-laminated onto the previous layer, where it is then laser-cut. The process is repeated until the object is formed. The process is inexpensive, but the finish level is low.

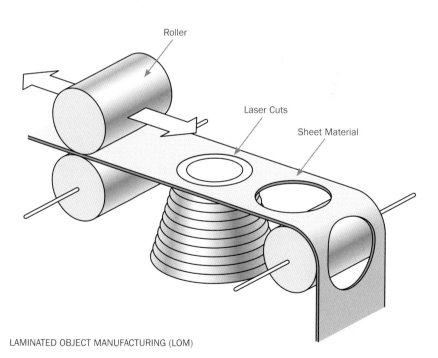

Roller

Laser Cuts

Sheet Material

LAMINATED OBJECT MANUFACTURING (LOM)

**Stereo Lithography (SLA)** SLA prototypes are progressively constructed from thin layers of liquid photopolymer that is selectively cured using an ultraviolet laser. A build platform is positioned just below the surface of the photopolymer. Then, using a 3D CAD data file, a scanning system is used to draw the first cross section layer on the surface of the photopolymer, which adheres to the platform. When the layer is complete, the platform moves lower into the vat of photopolymer and the next layer is drawn and cured, with each new layer adhering to the previous one. The process is repeated until the object is completed. Actual build times can range from less than an hour to more than a day, depending on the photopolymer, laser power, and complexity of the object geometry. Typically, a mechanical blade is used to sweep the surface of the photopolymer layer to ensure an even layer of resin for the next layer. This process typically yields highly detailed parts; however, secondary finishing processes are usually required.

**Selective Laser Sintering (SLS)** A process that is very similar to SLA; however, this system uses powdered material rather than liquid. It also allows for a variety of materials to be used, such as plastic, rubber, and metal. Because the powdered material supports the structure as it is being built, additional support structures are not needed, and undercuts are possible.

**Urethane castings** Components are produced using low-cost silicone or RTV (Room Temperature Vulcanization) molds. The process begins with a master pattern, which can be constructed using a variety of rapid prototyping or manufacturing techniques. Depending on the material to be cast, as well as other factors, the master pattern may be scaled to compensate for shrinkage in the final material. In most cases, draft is not an issue and small undercuts are possible because the silicone mold is flexible. Cast urethane parts closely simulate final injection-molded plastic parts. The quality and finish of the final parts are identical to the level of finish of the master part. Urethane castings can be molded in a selected color or painted.

Air vents

Two-part Mold

Final Parts

URETHANE CASTINGS

Courtesy of Scott Models

# Chapter 4: Material Selection

by Daniel F. Cuffaro

## OVERVIEW

Material selection is directly related to product performance requirements, cost, and user needs. The selection of manufacturing processes and materials is interrelated. However, within a manufacturing process there are usually many available choices of specifically formulated materials. This chapter provides an overview of some of the manufacturing processes and materials that are important to designers.

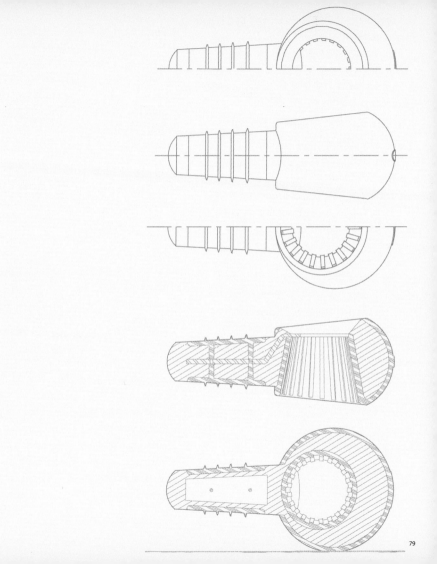

# GLOSSARY OF MATERIALS

**Ceramics** traditionally refers to clay, but more broadly refers to any nonmetallic inorganic (not plant or animal) material. Ceramic materials are strong and light but can be brittle. There are many types of clay, each offering different levels of refinement, colors, and performance characteristics. They are used in a broad range of applications from the more familiar consumer applications such as dinnerware and plumbing products (sinks, toilets) to high-tech and commercial applications such as semiconductors, electrical insulators, and building materials (bricks).

**Composites** are engineered materials made from two or more components. One component is a strong fiber that gives the material its tensile strength, while another component, called a matrix, is often a resin that binds the fibers together. Composites are used for their light weight and high strength, although they are often expensive. The definition of composites includes a broad range of materials, including carbon fiber, glass-fiber reinforced plastic (fiberglass), graphite-reinforced plastic, and Kevlar. Composites are used in applications related to automotive, aerospace, cycling, boating, and consumer products.

**Earthenware** is a low-fired clay (1975°F [1100°C]). The clay is usually red or orange and is usually somewhat porous. Similar to terra-cotta, earthenware is usually made from a finer clay that contains less grog.

**Elastomers,** often referred to as rubber, are characterized by their relative flexibility. Elastomers are usually thermoset materials (do not melt) but may also be thermoplastic (heat softens or liquefies). Elastomers are used extensively in consumer and commercial applications.

**Glass** is a transparent, relatively strong, durable, inert material that can be formed with smooth and impervious surfaces. Glass can be heat treated or tempered (to increase durability), colored, or coated. Glass is used extensively in consumer and commercial applications.

**Grog** is a clay that has been fired and then ground into fine granules. Grog imparts texture, opens the body to help uniform drying, and cuts down shrinkage in clay bodies. Between twenty and forty percent of grog may be used in a body depending upon the amount of detail desired and whether the pieces are freestanding or made in press molds.

**Metal** Specific metals are defined in terms of physical and mechanical properties. They tend to be strong, rigid, ductile, conductive (heat and electricity), opaque, and reflective. Generally, they are categorized as either ferrous (containing iron) or nonferrous. In order to achieve specific properties, metals are formulated into alloys, which are compositions of two or more chemical elements, with at least one being metal.

**Porcelain** is a vitrified, white, and translucent ware fired at (2400°F [1300°C]). At this temperature, the body and glaze mature together to create a thick body glaze layer. Porcelain has a refined look and feel.

**Stoneware** is a high-fired clay with slight or no moisture absorbency. It is usually gray to brown in color. Stoneware is similar in many respects to porcelain, the chief difference being the color, which is the result of iron and other impurities in the clay.

**Terra-cotta** has an earthenware body that is red in color and contains grog. It is often used in large sculpture and architectural forms.

## Ferrous Alloys

Ferrous alloys are created by burning carbon out of iron and adding other alloy elements. The addition of alloys and temperature treatments results in metals with increased strength, toughness, impact resistance, ductility, and other properties.

| | |
|---|---|
| **Cast Iron** | hard, brittle, and rusts easily; used in engine blocks and cooking pans |
| **Carbon Steel** | hard and strong, with low ductility; used in construction and machinery |
| **Alloy Steel** | formulated for specific strength characteristics; used in bridges, railroad cars, and construction vehicles |
| **Stainless Steel** | malleable and rust resistant; used in cutlery, architecture finishes, and automotive components |
| **Tool Steel** | very hard and brittle; used for drill bits, cutting tools, and tooling in manufacturing |
| **High-Strength** | high tensile strength and resists corrosion/abrasion; |
| **Low-Alloy Steel (HSLA)** | used in bridges, cranes, and vehicles |
| **High-Strength Steel** | high strength and light weight; used in construction applications |
| **Iron-Based Superalloys** | withstand high heat, wear, corrosion, and are creep-resistant; used in high-heat applications |

## Nonferrous Metals and Alloys

Nonferrous metals are well suited for reduced-weight/high-strength structural applications. They typically have lower melting points and are resistant to corrosion.

| | |
|---|---|
| **Aluminum** | strong, light, nonmetallic, and ductile; extensive consumer and commercial uses |
| **Beryllium** | strong, lightweight, and brittle; primarily; used as a hardening agent in alloys and in aerospace applications |
| **Copper** | malleable, ductile, and very conductive (heat and electricity); used in plumbing, electrical wiring, and consumer and commercial applications |
| **Lead** | dense, ductile, highly malleable, and poisonous; used in car batteries, fishing sinkers, ballast, and shot |
| **Magnesium** | strong and light; used in alloys for auto and aerospace applications |
| **Nickel** | hard, malleable, ductile, magnetic, and inert to oxidation; used in coinage, electroplating, and alloys |
| **Precious Metals** | gold is very soft, dense, malleable, ductile, and conductive; used for jewelry, coinage, and electronics<br><br>silver is the metal with the highest conductivity and a high degree of polish; used for jewelry, coinage, and electronics |
| **Refractory Metals** | Tungsten, Molybdenum, Niobium, Tantalum, Rhenium are very resistant to heat, wear, and corrosion; refractory metals are used in lighting, tools, and lubricants |
| **Tin** | malleable, ductile, and resists corrosion; used as a coating (tin cans) and for soldering |

| Titanium | very high strength-to-weight ratio and is corrosion resistant; used for aircraft parts, bicycles, golf clubs, and laptop computers |
|---|---|
| Zinc | moderately reactive to oxygen and other nonmetals; used in coatings (galvanized metals), battery containers, and paint pigments |
| Zirconium | light and corrosion resistant; used in lotions, piping, and heat exchangers |

## New Materials

**MEMS (Microelectromechanical Systems)** are microscopic electrical and mechanical devices built using processes similar to those used in silicon microchip fabrication. MEMS are used for electronics and computers and often are implemented on or used as microchips. This is a new technology with a vast potential for medical, aerospace, and commercial uses.

**Nanotechnology** is molecular manufacturing. It involves building objects one atom or molecule at a time with nanoscopic machines. It is believed that nanotechnologies will revolutionize the properties of materials and product performance (a nanometer is one billionth of a meter or four atoms wide).

**Plastics** are typically easy to form, are lightweight, and resist corrosion, though their properties vary widely. The also resist abrasion and yield a high finish level when formed. They are categorized as thermoplastic (or thermoform), which soften and melt when heated, or thermoset, which do not melt when heated.

## Thermoplastic

| | |
|---|---|
| **ABS (Acrylanitrile Butadiene Styrene)** | high impact and mechanical strength; used in durable consumer and industrial products |
| **ABS/PC (Polycarbonate)** | impact resistant, stiff, and strong; used in phones, small appliances, and computer enclosures |
| **PS (Polystyrene)** | rigid, dimensionally stable, brittle; used in signage, disposable dinnerware, and packaging |
| **PP (Polypropylene)** | rigid and chemical resistant; used in packaging, toys, appliances, and automotive components; mostly used to form living hinges |
| **PET Polyester (Polyethylene Terepthalate)** | hard, strong, and durable; used in soda bottles, textiles, water bottles and magnetic tape (audio, video) |
| **ACL (Acetal)** | rigid but not brittle, has a high melting point, and resistant to fatigue; used for moving parts in appliances, business machines, automotive door handles, seat belt components, plumbing fixtures, shaver cartridges, zippers, and gas tank caps |
| **Acrylic** | hard, rigid, transparent, and resistant to outdoor weathering; used for lighting diffusers, outdoor signs, automobile lights, and safety shields |
| **ASA (Acrylic-Styrene-Acrylonitrile)** | excellent outdoor weatherability and impact resistance; used in automotive body panels, outdoor products |
| **CAB (Cellulose Acetate Butyrate)** | strong, durable, hard, and clear; used in toys and sporting goods |
| **ETFE (Ethylene Tetrafluoroethylene)** | excellent chemical, thermal, and electrical properties, and is abrasion resistant; used in food packaging, linings and seals |

| | |
|---|---|
| **EVA (Ethylene Vinyl Acetate)** | flexible, transparent, and chemical resistant; used in vacuum cleaner hose and hand grips |
| **LCP (Polyester Liquid Crystal Polymer)** | very strong, resistant to high heat and chemicals; used in high-heat switches/sensors and cookware |
| **Nylon** | strong, elastic, chemical and water resistant, and has a high melting point; used in airbags, belts, hoses, tires, luggage, and packaging |
| **Nylon 4-6** | high-temperature and low-warp characteristics; used for high-temper ture and metal-replacement applications |
| **Nylon 6-6** | tough, has high fatigue stress, abrasion and impact resistant; used for moving parts (gears, latches, cams) |
| **Nylon 11** | high impact strength and low moisture absorption; used in automotive, medical, and electrical applications |
| **Nylon 12** | weather and UV resistant; used in automotive, medical, and domestic appliance applications |
| **Ethylene-Propylene Crystal Polymer** | sress-crack resistant, has lower temperature toughness, and is fatigue resistant; used in wire covering, film, and blow-molded bottles |
| **PC (Polycarbonate)** | strong, durable, and rigid; used in water bottles, appliances, and electrical components |
| **PEEK Polyetheretherkeytone)** | chemical, temperature, and electrical resistant; used in medical implants, automotive, and aerospace components |
| **PEI (Polyetherimide) Ultem** | strong, rigid, and heat, chemical, and flame resistant; used in electronics, aerospace, and medical applications |

continued

continued from page 85

| | |
|---|---|
| **PES (Polyethersulfone)** | dimensionally stable, has high weatherability, and is solvent resistant; used in medical products and electrical and structural components |
| **HDPE (High Density Polyethylene)** | chemical, fatigue, and wear resistant; used in containers |
| **LDPE (Low Density Polyethylene)** | flexible and moisture and corrosion resistant; used in lids and other household products |
| **PI (Polyimide) Aurum** | flame, chemical, and radiation resistant; used for gears, seals, fasteners, and wire insulation |
| **PPO (Polyphenylene Oxide) Noryl** | rigid, opaque, and durable; used in coffee pots, washing machines, and microwave parts |
| **PPS (Polyphenylene Sulfide)** | very stiff and chemical and heat resistant; |
| **HIPS (High Impact Polystyrene)** | hard, rigid, and translucent; used in refrigerator linings, toilet seats, and tanks |
| **PSO (Polysulfone)** | clear, strong, nontoxic, and chemical and heat resistant; used in medical, electronic, and automotive applications |
| **PVC (Polyvinyl Chloride)** | hard and brittle unless plasticizers are added; used for pipes (plumbing), upholstery (vinyl), and siding |
| **PVDF (Polyvinylidene Difluoride) Kynar** | expensive, pure, strong, and chemical resistant used in chemical, semiconductor, and medical industries, and for batteries and sensors |
| **SAN (Styrene Acrylonitrile)** | heat and chemical resistant; used in automobile interior components and toothbrushes |

## Thermoset Plastics

| | |
|---|---|
| **Alkyd** | heat resistant and dimensionally stable under high temperatures; used in ele trical components, automotive parts, and coatings |
| **Allylics** | hard, transparent, abrasion resistant, electrical insulators; used in optical coa ings, face shields, and electrical components |
| **Epoxy** | rigid, clear, and durable; used in protective coatings, adhesives, for encapsulating electrical components, and for impregnating fibers (fiberglass) |
| **Melamine** | very hard, durable, and colorable; used for dinnerware and in coating and adhesives |
| **Urea-Formaldehyde** | very hard, scratch, heat, and chemical resistant; used in laminates, coatings, decorative, and electrical products |
| **Phenolic** | hard, brittle, and opaque; used for appliance knobs and handles and in electrical components |
| **Polyester** | good mechanical properties, and is heat resistant; used in textiles, bottles, films, composites, and liquid crystal displays |
| **Polyurethane** | in foamed state may be flexible or rigid, excellent insulator; used in cushions, padding, building, and appliance insulation |

# Chapter 5: Electronics

by Larry Sears

## OVERVIEW

Given the degree to which electronics have saturated our products and our lives, designers must be equipped to understand the basic principles of electricity and their applications. This chapter will not provide the necessary knowledge to create complex electrical circuits but will provide enough information so that designers can make intelligent decisions about design approaches involving electricity and electrical devices.

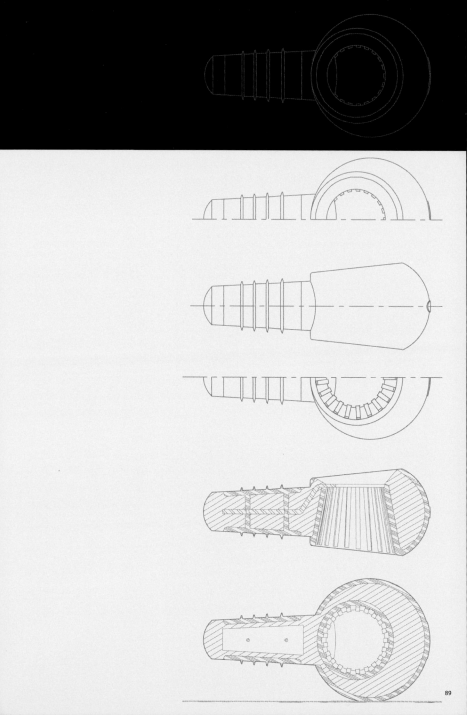

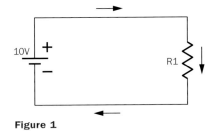

**Figure 1**

## Volts, Amps, Ohms, and Circuits

The fundamental element of an electrical circuit is a voltage source. A voltage source, when connected to an electrical circuit, establishes a flow of electrons through the circuit. An example of a simple circuit is shown above.

In this circuit a battery is the voltage source, and it is connected in series with a resistor R1 (often known as the load). The resistor is an element that restricts the flow of electrons. The circuit includes wires to connect the circuit elements. Note that the wires (usually copper) are conductors of electricity and readily permit the flow of electrons. Conductors have some resistance, but their resistance is so low that it can usually be ignored. (Conductors are poor resistors, and resistors are poor conductors).

In figure 1, the voltage source establishes a current (I) through the battery, conductors, and resistor. By convention, the current flow is shown to flow from the positive side of the source to the negative, as shown. The magnitude of the current is determined by the potential of the battery and the value of the resistor. The celebrated Ohm's Law represents this relationship:

$$I = V/R$$

From this equation, the current I (in amps) is equal to the source voltage divided by the value of the resistor (in the ohms). So, if a 10-volt battery is used as a source and the resistor value is 100 ohms, the current through the circuit will be 0.1 amp.

When the battery was manufactured, it contained stored energy in the form of a potential chemical reaction. When the battery is connected to a circuit, the current that flows from the battery does work that gradually expends this energy. In the case of figure 1, the resistor produces heat (in watts) as energy is slowly transferred from the battery to the resistor. The following equation calculates the power dissipated in the resistor:

$$P = I^2R$$

In this example, $P = (0.1)^2 \times 100$, or 1.0 watt.

We now have the two fundamental equations that determine the operating parameters of a circuit. With simple algebra, we can substitute and rearrange the two equations to derive other relationships:

$$V = I \times R \qquad R = V/I \qquad P = V^2/R$$

To construct the circuit in figure 1, a resistor would have to be chosen so that it could safely dissipate 1.0 watt without overheating. To produce a suitable resistor, a manufacturer would typically provide a small cylinder with wire leads containing an electrically resistive material. The surface area of the resistor would have to be large enough so that the resistor will transfer enough heat to the surrounding air by radiation and conduction so it does not overheat. If a resistor were used in a circuit that was rated at 1.0 watt, the temperature of the resistor would stabilize at about 120°F (50°C) above ambient temperature.

If the resistor is replaced by an incandescent lamp, all the dissipation would be concentrated in a small filament. Since the surface area of the filament is very small, little heat is lost to the surroundings, and the filament will heat to the point of incandescence. If enough voltage and current were available, a resistance heater, as used in a toaster, could be constructed by using a long length of special resistive nichrome wire. In this case, the resistance, length, and surface area of the resistance wire are selected to produce the desired dissipation and temperature.

## Multielement Circuits

The circuit in figure 1 (see page 90) is the basic complete circuit. Figure 2 illustrates how additional elements can be combined.

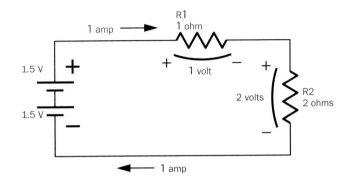

**Figure 2**

Note that current can never be "lost" in a circuit. Therefore, the current is the same throughout the loop. Similarly, voltage can never be lost. In this example the total voltage across the load resistors must equal the total source voltage

In this case the values of the two voltage sources combine to produce a voltage source of 3 volts. The resistors are in series, so their values add to produce a total load resistance of 3 ohms. Using Ohm's Law, the current through the circuit is 1 amp. Note that the two resistors comprise a **voltage divider** with 2 V appearing across R2 and 1 V appearing across R1.

A common application of Ohm's Law is shown opposite (figure 3). A 9 V battery is used to drive a light-emitting diode (**LED**). From its data sheet, the LED requires a current of 20 milliamps (0.02 amp) and it will have a voltage drop of about 1.8 V. The total voltage across the resistor is 9 V – 1.8 V, or 7.2 V. From Ohm's Law, we select R = 7.2/0.02 = 360 ohms. The power dissipated in the resistor will be 0.14 watt. Resistors are only available in certain discrete values, so a 390 ohm, 1/4 watt part would be selected.

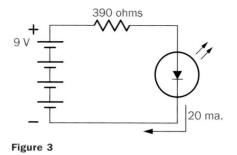

**Figure 3**

## Switches and Contacts

It is usually necessary to interrupt the current flow in a circuit. A switch is a simple device that consists of a conductor that can be opened and closed to start and stop the flow of current. This simple device can, however, be quite unreliable because conductors can be in physical contact but may not provide a low resistance path. This is due to oxides and insulating films that can form on metal surfaces and prevent electrical contact. This can be demonstrated by striking a flashlight on a table to improve its brightness; the battery motion breaks the insulating layer. Reliable switches employ contacts that close with suitable force, have a wiping or sliding action to provide a clean contact surface, and are plated to resist corrosion and oxide films.

## Batteries

The key battery parameter is **capacity**, which is specified in **amp-hours**. The A-H rating specifies the current that a battery can provide for a given time before its voltage drops below a specified value. For example, an AA alkaline battery has a rating of about 1.5 A-H and can provide 0.1 amp for 15 hours before its voltage drops to 1.0 V. The same battery could also provide 0.2 amp for about 7.5 hours before its voltage drops to the same value. The A-H rating of batteries will vary over a wide range depending on the discharge conditions. For example, the capacity of an AA alkaline cell drops to 0.5 A-H if the load is increased to 1.0 amp. Realizable capacity also varies depending on temperature, battery quality, discharge duty cycle, and end-of-life voltage

Batteries are available in many types and sizes, including rechargeable and disposable styles. Disposable batteries include common alkaline flashlight cells, cylindrical lithium cells, and various types of "button" cells. Rechargeable cells include lead-acid automobile batteries and nickel-cadmium and nickel metal-hydride batteries as used in power tools. The following table provides an overview of the characteristics of different cells.

| Cell Type | Cell Cost | Energy Density | Voltage Stability | Shelf Life | Charging Circuit Complexity |
|---|---|---|---|---|---|
| | | capacity vs. size and weight | maintenance of output voltage as cell discharges | capacity loss through self discharge | |
| Alkaline | Low | High | Poor | Good | N/A |
| Lead-acid | Medium | Low | Fair | Poor | Medium (fast charge) |
| Ni-Cad | Medium | Medium | Good | Fair | Low (slow charge) |
| NiMH | High | High | Good | Fair | High (fast charge) |

## Alternating Current

Most electronic devices operate on low-voltage (1.5 to 24 V) DC supplies. Power distribution, however, utilizes high-voltage alternating current. Alternating current (AC) voltage sources provide a sinusoidal waveform that alternates in polarity. In the U.S. the standard residential 120 VAC line varies between peaks of -170 and +170 V, and one complete cycle occurs 60 times per second. When this waveform is applied to a resistance heater, the dissipated power is equivalent to that of a DC source with a value of 120 V. This is known as the RMS voltage, which is the standard way of describing the value of a sinusoidal waveform.

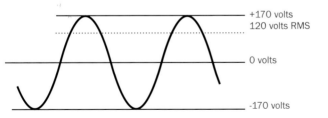

+170 volts
120 volts RMS

0 volts

-170 volts

With a sinusoidal waveform, the positive and negative peaks are always equal to 1.4 times the RMS value.

**Figure 4**

In most countries outside of North America, the residential line voltage is 220 VAC, 50 Hz. Japan uses 100 VAC, 50 or 60 Hz, depending on the region.

## POWER DISTRIBUTION AND SAFETY

A simplified diagram of a U.S. residential power system is given in figure 5A.

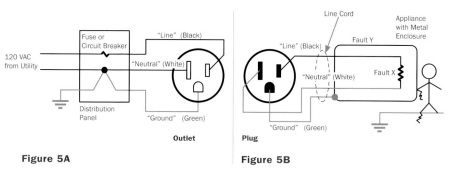

**Figure 5A**

**Figure 5B**

A three-wire grounded outlet is always wired with black, white, and green (or bare) wires. At the distribution panel (near the utility meter) the black wire is connected to one side of the voltage source (known as line) through a fuse or circuit breaker; the white wire is connected to the other side of the source (known as neutral) as well as to a metal pipe or other type of ground to earth. The green wire is a "safety ground" and is connected to the outlet ground terminal (the round one) and connected back to the grounding point.

In figure 5B an electrical appliance with a metal housing is shown plugged into the outlet. Current flows to the load through the white and black wires. The green wire connects only to the metal alliance housing and does not connect to any part of the circuit; no current flows in the green wire.

Note that the appliance housing is directly connected to earth ground via the green safety wire. Should there be a short circuit across the load (shown as a wire labeled fault X), the current in the black and white wires would increase to a very great value and the fuse or circuit breaker would blow; there would be no danger to the operator in this case, and the open circuit breaker would terminate the flow of dangerous current. However, if there is a fault, such as a frayed line cord, that connects the line to the metal enclosure (fault Y in figure 5B), the full 120 V would appear between the enclosure and ground. If the operator were to touch the enclosure while in contact with the ground (through wet earth or by touching any grounded metal object, for example), the operator's body would complete the circuit from line to neutral and could receive a severe shock. However, if the enclosure is properly connected to the grounded terminal of the outlet, the fault current would pass through the low-resistance safety ground instead of through the operator's body. This low resistance ground circuit would protect the operator, and at the same time the heavy fault current would blow the fuse or circuit breaker and remove the voltage from the outlet.

It is imperative that precautions are taken with the design of products such as tools that may be used outdoors, or in any other case when the operator is likely to be grounded. Large appliances usually employ a three-wire cord and plug for use with a grounded outlet; this provides the means for grounding the enclosure or other exposed metal parts that may inadvertently come in contact with the high side of the AC line. In some cases, a double-insulated design is used. This configuration employs high-quality redundant insulation to prevent the operator from contacting the AC line. Double-insulated equipment can be safely used with older household wiring that has "two-wire" outlets instead of newer, grounded, three-wire outlets.

Many electrical devices, such as table lamps, use two-wire cords and do not have double insulation. These devices can be quite hazardous, even though they are accepted in the United States for use in locations where it is unlikely that the operator would be grounded.

---

**Safety Approval Agencies**

In an effort to enhance product safety, insurance companies founded organizations that establish safety requirements and test new products for compliance. Agencies such as Underwriter's Laboratories publish detailed standards for almost every type of product and issue a certificate of approval when they have thoroughly and successfully tested samples of a device. Note that these agencies are only concerned with product safety. Product performance or reliability is generally not evaluated.

## Electromagnets

When current is passed through a coil of wire surrounding a core of ferrous material, a magnetic field is established in the core and an electromagnet results. Electromagnets can be used in a number of ways to produce devices that convert electric current to mechanical motion. In all cases, the most efficient devices result when the motion of the device provides a complete path for the magnetic field within the core material.

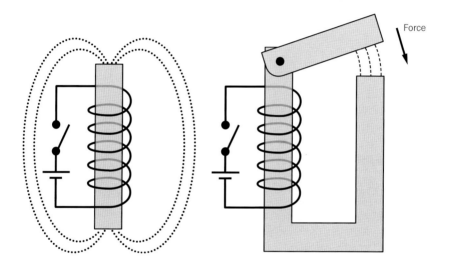

| Figure 6A | Figure 6B |

In figure 6A, the current in the coil is determined by the battery voltage and the resistance of the wire in the coil. This current will produce the magnetic field shown. If an iron bar is added, as in figure 6B, the flux will be constrained in the iron and concentrated in the air gap. When the coil is energized, the free end of the bar will be strongly attracted to the core in an effort to close the gap and complete the magnetic circuit. As long as the gap is small (typically less than 0.1 inch [2 mm]), the attractive force will be large and useful work can be performed.

This type of actuator may be used to strike a bell or release a latch. Another example of this configuration is a relay, as shown in figure 7, which permits a small coil current to open or close a set of contacts. The contacts can switch high voltages and currents or can be used to provide simple logic functions. A solenoid is a cylindrical version of a magnetic actuator that has a plunger that can provide an axial pushing or pulling force.

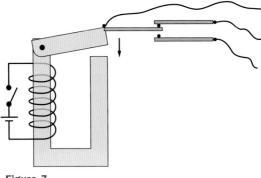

**Figure 7**

## Electric Motors

Electric motors produce rotary motion using electromagnetic force. An output shaft is attached to a rotating armature containing electromagnets. The armature is caused to rotate as it continuously reacts with a magnetic field established in the surrounding stator containing stationary magnets (either permanent magnets or electromagnets). In the case of the magnetic actuators discussed above, the moving iron bar closed the gap and its motion stopped. However, with motors there is a means of rotating or switching the magnetic field so that the armature keeps chasing the field but never catches it. Some common motor types are discussed below.

DC permanent magnet motors provide excellent performance at exceptionally low cost, and are used in products ranging from tiny toys to hand-held power tools. These motors use permanent magnets in the stator and a commutator to switch electromagnets in the armature. The commutator consists of a ring of copper segments that is mounted on the shaft and is contacted by spring-loaded stationary brushes. The brushes connect to the power source and sequentially energize the armature electromagnets as the shaft rotates. One limitation on the application of these motors is that the brushes wear and lifetime can be very short.

The speed of a permanent magnet motor is proportional to voltage; therefore the speed of permanent magnet motors can be conveniently controlled by varying the input voltage. The output torque is proportional to the current drain, but since motors have resistance in their magnet coils, as the shaft is loaded the in-creased current causes an apparent voltage drop at the motor. As a result, motor speed decreases with load; this response is specified by the motor's speed-torque curve. Note that a motor that is starting will briefly draw very high current since much torque is initially required to overcome inertia.

**AC induction motors** A class of motors that utilize AC power to create a rotating stator field that reacts with a rotor mounted on the output shaft. The field rotates at a rate that is fixed by the line frequency, so induction motors are inherently constant-speed motors. For a given size, AC motors have lower output power than a permanent magnet motor, and they are limited to applications such as fans that require little starting torque. However, they do not have brushes, so they are quieter and can have essentially unlimited life.

**Stepping motors** are extremely popular in applications such as printers where accurate motion control is required. In a stepping motor there is no commutator; in fact, the motor is not designed to smoothly rotate but to step from one position to another through a small angle. A stepping motor has a permanent magnet rotor, and the stator is a group of electromagnets that can be individually energized. By sequencing the stator coils, the motor will step from position to position as established by the angular spacing of the stator magnets. Stepping motors are available in a wide variety of sizes, and step angles range from under 1 degree to 15 degrees. Note that these motors are inherently low-speed devices (a few hundred rpm is a typical maximum), require an electronic drive circuitry, and can be extremely inefficient under most conditions. Permanent magnet motors are a far better choice if simple rotation is required.

**AC/DC motors** are sometime called universal motors and are appropriate for certain applications. These are brush-type motors that have a wound stator, wired in series with the brushes, instead of permanent magnets. These motors are generally operated from the AC line and have extremely high no-load speed that drops dramatically under load. These motors feature very high starting torque, high output, small size, and very low cost. They are used in many appliance applications, such as garbage disposals and blenders, where high power and low cost can be traded off against limited lifetime and noisy operation.

## Transformers

It was stated above that most solid-state circuits operate on low-voltage direct current. This is convenient for a battery-operated product since the battery can directly source low-voltage DC. However, if the product is to be operated from the AC line, it is cumbersome and inconvenient to provide low voltage. There are three reasons why it is difficult to operate low-voltage devices from the AC power line.

First, since the AC line is 120 V, a means is needed to drop the voltage. It is possible to reduce the voltage with a simple series voltage-dropping resistor, but since so much voltage must be dropped, unless the current drain is very small a large amount of power (and heat) must be dissipated in the resistor. Second, if a line-to-neutral short circuit should occur in the product, a fire hazard could result due to the significant current that can be drawn from the AC line. The third problem is also related to safety. If a product is line-operated, there will be significant requirements on the design to prevent the operator from coming in contact with the AC line. For example, a product such as a CD player would require heavy enclosure walls, elimination of all exposed metal parts, and a heavily insulated headphone and cord.

All of these problems can be eliminated by the use of a transformer. A transformer consists of two coils wrapped on one magnetic core, with no electrical connection between them. The primary coil contains a large number of turns and is energized by the AC line. The secondary coil contains a smaller number of turns and provides a reduced output voltage that is proportional to the input voltage and the turns ratio of the coils. There is little power loss in transformers, and, most importantly, the transformer secondary is electrically isolated from the AC line. This means that if any of the output leads of the transformer touched the operator or became grounded, no safety issues would result.

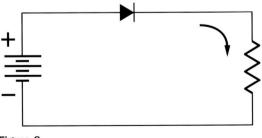

**Figure 8**

The use of small plug-in wall-mount transformers has become very popular. These devices are convenient because their size and weight may be removed from a portable product, and they may only be needed when charging batteries. These transformers are also designed to have limited output current; because of this feature a failure that short-circuits the transformer output will not cause excessive heat or current flow. Finally, wall-mount transformers are double insulated, so they can be safely used in products that do not have a safety ground.

## Diodes and Rectifiers

Whereas transformers provide isolated low voltage, it is still necessary to convert the AC voltage from the transformer secondary into DC voltage. To accomplish this, semiconductor diodes are employed, which allow current to flow in only one direction. In the circuit shown in figure 8, current will flow as indicated. However if the battery polarity is reversed, no current will flow. If the battery is replaced with an AC voltage source as shown in figure 9A, current will flow only for one half-cycle. A much steadier output will result if a bridge rectifier is used as in figure 9B.

The pulsating output from a half-wave or full-wave rectifier is suitable for many DC applications, including battery chargers, lamps, and motors. In other cases, such as those where electronic circuits are present, it is necessary to reduce this ripple voltage. This is accomplished with the addition of a filter capacitor that is placed across the output of the rectifier. The capacitor acts as a storage device (similar to a small battery) that is charged by the AC voltage source at the peak of each cycle; during this period, a relatively large current flows from the voltage source. When the voltage at the source drops below the capacitor voltage, the source current stops flowing due

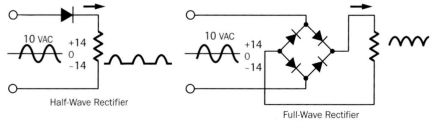

Half-Wave Rectifier

**Figure 9A**

Full-Wave Rectifier

**Figure 9B**

to the diode(s) and the load current is supplied solely by the capacitor. To provide a smooth output, the capacitance of the capacitor must be great enough to provide the load current during the relatively long intervals when the source current is zero. As a rough guide, a full-wave rectifier with a 1/2 amp load will have a ripple voltage of about 2 volts (measured "peak-to-peak") when a 1000 mfd filter capacitor is used. Filter capacitors are physically large and polarity sensitive. They also have a maximum voltage rating that must never be exceeded.

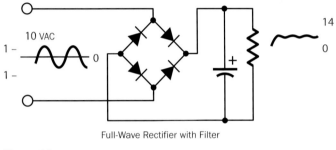

Full-Wave Rectifier with Filter

**Figure 10**

TYPES OF ELECTRICAL SOCKETS—WORLDWIDE

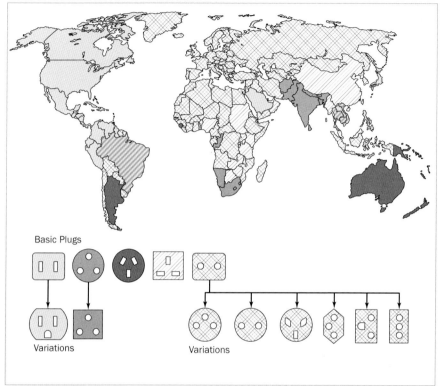

**Figure 11**

Worldwide, there are several voltage and plug architecture standards. When designing a product intended for a specific country or for the international marketplace, be sure to accommodate a variety of requirements.

# Chapter 6: Intellectual Property

by Darrell E. Covert

## OVERVIEW

According to The World Intellectual Property Organization, "Intellectual property (IP) refers to creations of the mind: inventions, literary and artistic works, and symbols, names, images, and designs used in commerce." When a designer creates a novel work, there are many issues to navigate relative to protecting and capitalizing from intellectual property rights. Creative works and inventions can be protected with patents, trademarks, or copyrights, and this chapter describes some of the options and processes for establishing property rights.

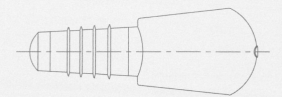

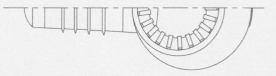

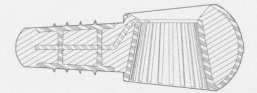

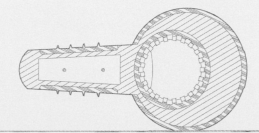

## TYPES OF IP PROTECTION

Presuming that one developed an idea or work that warrants IP protection, it is important to determine what form of protection is most appropriate for the work produced. Patents are particularly important in product development. However, the following represents a brief description of IP protection options:

**Patent** A patent offers protection for inventions and improvements to existing inventions. While the scope of what is patentable is broad, two examples related to product development include 1) a surgical stapler with a novel cutting and stapling mechanism that is more effective and less expensive (utility patent), and 2) a surgical stapler with a unique brand-identifiable shape and appearance (design patent).

**Trademark** A trademark offers protection for any word or visual symbol that is linked to a product or service to distinguish that product in the marketplace. Two examples of trademarks include 1) the specific name of an automotive brand, and 2) the specific logo of an automotive brand.

**Copyright** A copyright offers protection for original creative expressions of ideas unique to an individual, distinct to other prior existing interpretations in literary, musical, dramatic, or artistic concepts. Two examples of copyrights include 1) a product development book, and 2) a specific chart or graph within a book.

**Distinguishing between patents, trademarks, and copyrights** Patents, trademarks, and copyrights are similar in that they protect individual rights of ownership to creative works. However, patents are typically focused on mechanical devices, mechanisms, or unique processes, or appearances related to these products. Copyrights and trademarks protect graphics, images, words, logos, or unique or nonproduct creative works.

### Patent Eligibility

Patent law describes the types of things that can be patented and the conditions with which they are eligible. Anyone who invents or discovers any new and useful process, machine, manufacture, or composition of matter, or any new and useful improvement thereof, may obtain a patent, subject to the conditions and requirements of the law. The laws of nature, physical phenomena, and abstract ideas are not patentable, and a patent cannot be obtained upon a mere idea or suggestion. It further states that an invention or design must be novel (different from all other existing products) and unobvious (creates unexpected results).

## Patent Distinctions

**1. Utility** A utility patent may be granted to anyone who invents or discovers any new and useful process, machine, article of manufacture, or composition of matter, or any new and useful improvement thereof.

**2. Design** A design patent may be granted to anyone who invents a new, original, and ornamental design for an article of manufacture.

**3. Distinguishing between a utility and design patent** If changing the shape of a design changes the function or performance of the design, then it can be considered for a utility patent.

## Provisional Patents

Provisional patents are a means of providing protection to a design idea, establishing an early effective filing date. It allows for filing without a formal patent claim, oath or declaration, or any information disclosure, (prior art) statement, and automatically becomes abandoned after one year.

Usually, provisional patents are applied for while completion of full patent documentation is taking place. Another benefit of using provisional patents is to conduct parallel activities such as determining a product's full market potential, or seeking the sale or licensing of the intellectual property.

It also allows the term *patent pending* to be applied—a phrase that often appears on manufactured items. It means that someone has applied for a patent on an invention that is contained in the manufactured item. It serves as a warning that a patent may be issued that would cover the item and that copiers should be careful because they might infringe if the patent is issued. Once the patent is issued, the patent owner will stop using the phrase *patent pending* and start using a phrase such as "covered by U.S. Patent Number XXXXXXX." Applying the *patent pending phrase* to an item when no patent application has been made can result in a fine.

PATENT DECISION-MAKING CHART

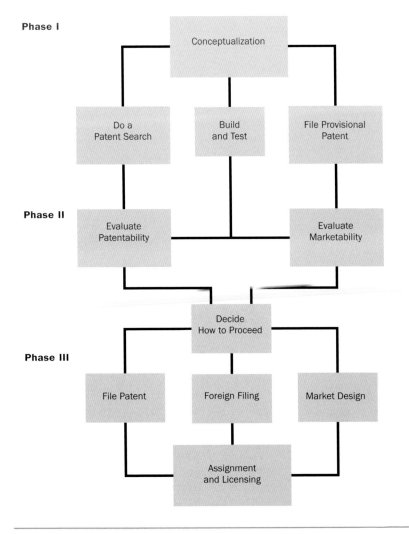

**Phase I**    Conceptualization

Do a Patent Search    Build and Test    File Provisional Patent

**Phase II**    Evaluate Patentability    Evaluate Marketability

Decide How to Proceed

**Phase III**    File Patent    Foreign Filing    Market Design

Assignment and Licensing

## Evaluating the Value of a Design

Fewer than one patented invention in ten generates a profit for the patent holder because of poor marketing and feasibility testing. An invention must produce a minimum of $50,000 USD in sales to justify patent costs and expenses. It is important that the financial opportunity be thoroughly evaluated before making an investment in the process. Many factors should be considered, such as competitive products, barriers to entering the marketplace, the economy, and the advice of experts.

When considering how to proceed with IP protection, keep in mind all the options, including discontinuing the process if the outcome is not promising, selling the idea to a manufacturer (perhaps having only a provisional patent), or filing for a patent and bringing the product to market. The chart below is useful guide to decision making.

DESIGN DECISION MAKING CHART

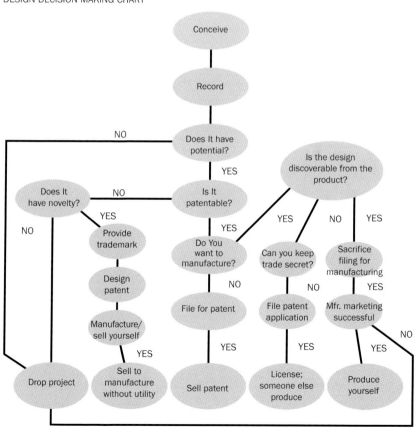

## Patent Searches

A patentability search is an attempt to determine whether an invention has merit to be awarded a patent. Inventors and designers need to search for existing patents in their fields to validate their creations and discoveries in the context of prior art or to modify their creation to merit the award of a patent. The process of the search is a crucial step that needs to be fully explored within product development before any actual design work can be completed and documented. This practice could avoid surprises that could increase the cost of a development program as well as delay its completion.

## Why Make a Search?

- To better understand what has already been created
- To discover if the design may infringe on existing work
- To determine if the design has already been patented
- To avoid unnecessary work that may already be claimed
- To avoid describing components that may already exist
- To be able to design around existing work and avoid future conflict
- To provide documented justification for the "uncommon solution" criteria
- To provide background on the patent application
- To gain additional insight into competitors' designs
- To discover what is novel to better exploit its future value
- To provide the best possible patent protection
- To expedite the patent application examination

## Ways to Search for Patents

### 1. Do it yourself:

• Search the patent database under different criteria, finding patent descriptions as well as drawings. Information in the United States is available through the USPTO website, www.uspto.gov/patft/index.html. Many patent office websites outside the U.S. have similar information.

• In the United States visit and search the USPTO in Alexandria, VA.

• Search a local Patent and Trademark Depository Library.

• Search stores, catalogues, reference books, and product directories.

• For patents outside the U.S. you can find information at http://ep.espacenet.com (an extensive list of resources can also be found at the end of this chapter).

### 2. Seek assistance:

• State planning and development agencies or departments of commerce and industry actively seek new product and new process ideas to assist manufacturers and communities in the state. For information, contact the state governor's office.

• Patent agents and patent attorneys can be found in the USPTO publication "Attorneys and Agents Registered to Practice before the U.S. Patent and Trademark Office." For a copy, visit the USPTO website.

# PREPARING YOUR DESIGN FOR PATENT APPLICATION

### Proving Ownership

- The best way to prove creative ownership is with a bound, sequentially numbered notebook where there is a recorded history of inventions as they have taken place, with your signature, witnessed and dated by others.

- Sketches, drawings, computer prints, or other material too large to fit can be reproduced and mounted to pages in the notebook to be signed and sequentially numbered.

- Prototypes, models, and other preliminary products may be photographed for ease of documentation. The photographs or copies may be sequentially mounted and recorded to maintain history with each page signed and dated.

### Application Content Guidelines

- The description of the invention should be detailed so that one could make or use the design based on the material provided.

- The major design claims should be as broad as possible to allow maximum protection with the patent.

- Key features and unique advantages of the invention should be highlighted in the application.

### Patent Drawing Requirements

The application for a patent requires that a drawing of the invention be furnished when appropriate. The USPTO has very specific guidelines for patent drawings that result in uniform, easy to understand representation of the invention for anyone using patent descriptions.

### Computer Generated Drawings

With the availability today of multiple inexpensive drafting software programs, it becomes a relatively easy matter to create an isometric view or exploded view of your design idea to be used in your patent discloser. However, understand that the Patent Office has some very specific requirements on how it wants these drawings created and presented for patent application.

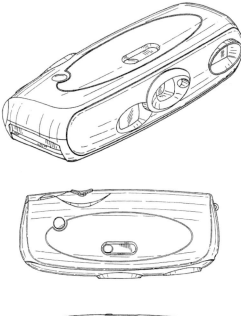

As with anything that requires a specialist, it may be best for you to consult someone specializing in the creation of patent drawings to take your design drawings, computer generated or otherwise, and put them in a final format that will conform to the current requirements of the patent drawing requirements.

Patent drawings all share a similar style, which is intended to universally highlight functional and aesthetic characteristics of the invention. Multiple views are used to thoroughly communicate the design.

**The European Patent Office (EPO)** in Munich, Germany, offers the distinct advantage of full European patent coverage. By filing your patent with the EPO you can elect to register in multiple European countries without additional submittals. This significantly reduces the difficulty associated with preparing multiple filings and dealing with country to country regulations.

## PATENT RESOURCES

### Internet Resources

- **Trademark.com**
  Provider of online trademark searches, registration, and renewal services.
  www.4trademark.com

- **Canadian Intellectual Property Office**
  Responsible for the administration and processing of intellectual property in Canada.
  www.opic.gc.ca

- **Canadian Patent Bibliographic Database**
  Provides access to more than seventy-five years of patent descriptions and images, and more than 1.4 million Canadian patent documents.
  www.patents1.ic.gc.ca

- **European Patent Organization (EPO)**
  The European Patent Organization establishes a uniform patent system in Europe for all contracting states and oversees the protection of intellectual property throughout Eurasia. A list of patent-related websites is maintained by the European Patent Office.
  www.epo.org

- **European Office for Harmonization of the Internal Market**
  Registers community trademarks and designs.
  www.oami.eu.int

- **France: National Institute for Industrial Property**
  Official site of the French Institute for Industrial Property.
  www.inpi.fr

- **Germany: Patent and Trademark Office**
  Official site of the German Patent and Trademark Office.
  www.dpma.de

- **Hong Kong Special Administrative Region Government: Intellectual Property Department**
  www.info.gov.hk

- **Intellectual Property (IP) Australia**
  Official website for IP Australia. The federal government agency that grants rights in patents, trademarks, and designs.
  www.ipaustralia.gov.au

- **Intellectual Property Department Hong Kong Government**
  Works toward the protection of intellectual property rights and provides patent, trademarks, and design registration services to the public in Hong Kong.
  www.info.gov.hk

- **Intellectual Property Rights Helpdesk**
  Free service supporting creativity and innovation in Europe. A project of the European Commission DG Enterprise.
  www.cordis.lu

- **Official Gazette Notices (1964–1996)**
  Search engine for accessing official gazette notices provided by the U.S. Patent and Trademark Office.
  www.uspto.gov

- **Trilateral Cooperation**
  The European Patent Office (EPO), the Japanese Patent Office (JPO), and the U.S. Patent and Trademark Office (USPTO) exchange information and views regarding patent administration in general, patent documentation and classification, automation programs, and patent examination practice.
  www.european-patent-office.org

- **United Kingdom Patent Office**
  Helps to stimulate innovation and the international competitiveness of industry through intellectual property rights; patents, designs, trademarks, and copyright.
  www.patent.gov.uk

- **U.S. Patent Copyright Office**
  www.copyright.gov

- **United States Patent and Trademark Office**
  Catalogs all U.S. patents by subject matter and implements the patent process for inventors.
  www.uspto.gov/patft/index.html

- **WIPO PCT Gazette Search Site**
  Provides access to the World Intellectual Property Organization (WIPO) Patent Cooperation Treaty (PCT) Database.
  www.pctgazette.wipo.int

- **World Intellectual Property Organization (WIPO)**
  An intergovernmental organization headquartered in Geneva, Switzerland. This specialized agency of the United Nations is responsible for the promotion of the protection of intellectual property throughout the world through cooperation among states.
  www.wipo.int

## Patent and Trademark Office Phone Numbers in the United States

Patent Trademark Office
703-308-HELP

Initial Patent Examination
703-308-1202

Provisional Patent Application
800-786-9199

Pending Patent Application
703-308-2733

Patent File History
703-308-2733

Foreign Patents
703-308-1076

Patent Cooperative Treaty Office
703-308-3257

# Chapter 7: Ergonomics

by Carla J. Blackman

## OVERVIEW

If you ask several people what *ergonomics* is, you are likely to hear a variety of descriptions. You are also likely to hear it called human factors; the two terms are synonymous. The short definition that is easy to remember and explain to others in conversation is "the process of designing for human use."

The comprehensive definition from Chapanis (1985) is "Ergonomics discovers and applies information about human behavior, abilities, limitations, and other charac-teristics to the design of tools, machines, systems, tasks, jobs, and environments for productive, safe, comfortable, and effective human use."

Both definitions share that all-important element: human use. The user is the focal point of all industrial design, and the best products are those that respect the user... the installer, the owner, and the servicer.

When asked what is your process for designing, Niels Diffrient, the famous designer and author of *Humanscale*, replied, "I look at the ergonomics first, then consider the engineering; after that the aesthetics fall into place."

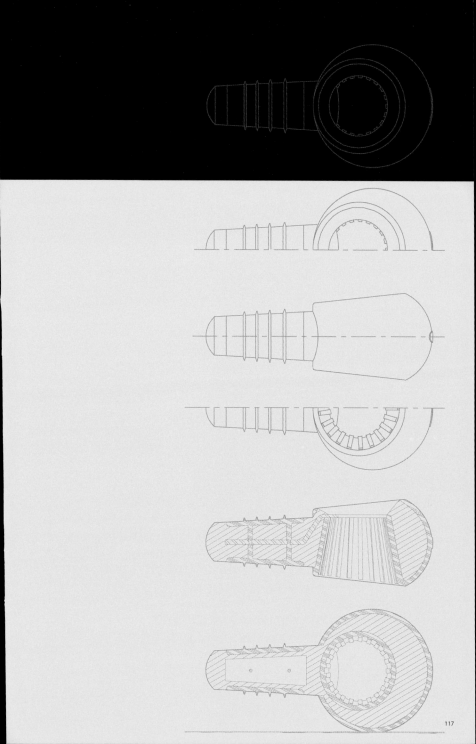

## Objectives and Approach

The essence of ergonomics is to develop a satisfying user experience. To help that process, the designer should keep in mind the objective: enhancing the effectiveness of physical objects and enhancing desirable human values of safety and human comfort. The approach can be described as systematic application of relevant information to the design process. In evaluating products, designers should ask of the product: Is it safe? Is it comfortable? Is it robust, reliable, convenient, and attractive?

---

# PERCENTILES AND ANTHROPOMETRICS

One important element of ergonomics is anthropometrics, which is a systematic collection and correlation of measurements of the human body. A good anthropometric survey includes males, females, children, and the elderly, and is large enough to statistically represent the entire population. Once a statistically significant body of measurements is complied, then percentiles can be understood. A percentile is a value that represents the percentage of people at or below a certain measurement. See graphic on pages 120-121.

Associating a typical user group with a specific design forces decisions on what percentile of the population to accommodate. A typical range of accomodation is from the 2.5 percentile female (only 2.5 percent of the female population is smaller) through the 97.5 percentile male (larger than 97.5 percent of men). More simplistic anthropometric surveys include a range from 50 percentile male to 50 percentile female, or the averages of these dimensions. It is important to accommodate a large sample of users, including those that represent different body types.

The importance of anthropometric data goes beyond the static human figure. Measurement of the body in motion and relative to specific tasks is also important.

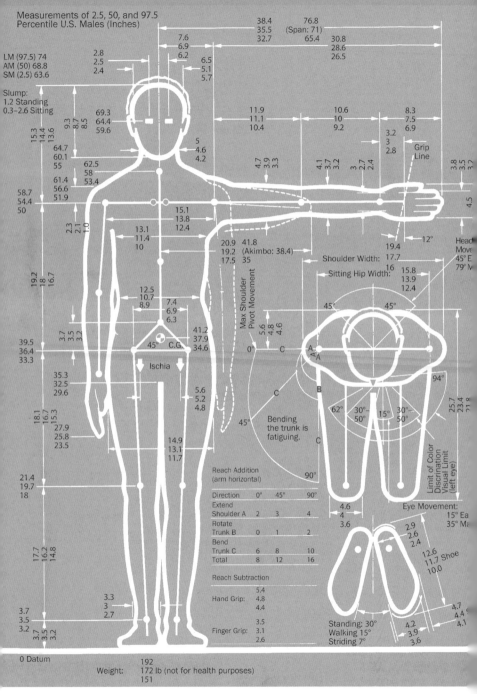

Measurements of 2.5, 50, and 97.5
Percentile U.S. Males (Inches)

38.4
35.5 (Span: 71)
32.7 65.4
76.8

30.8
28.6
26.5

LM (97.5) 74
AM (50) 68.8
SM (2.5) 63.6

7.6
6.9
6.2

6.5
5.1
5.7

2.8
2.5
2.4

Slump:
1.2 Standing
0.3–2.6 Sitting

69.3
64.4
59.6

9.3
8.7
8.5

15.3
14.4
13.6

11.9
11.1
10.4

10.6
10
9.2

8.3
7.5
6.9

3.2
3
2.8

Grip
Line

64.7
60.1
55

62.5
58
53.4

5
4.6
4.2

4.7
3.9
3.3

4.1
3.7
3.2

3
2.7
2.4

3.8
3.5
3.2

61.4
56.6
51.9

58.7
54.4
50

2.3
2.1
1.0

15.1
13.8
12.4

4.5

12°

Head
Move
45° E
79° M

19.4

Shoulder Width: 17.7

13.1
11.4
10

20.9 41.8
19.2 (Akimbo: 38.4)
17.5 35

16
15.8
13.9
12.4

Sitting Hip Width:

19.2
18
16.7

3.7
3.5
3.2

Max Shoulder
Pivot Movement

45° 45°

12.5
10.7
8.9

7.4
6.9
6.3

5.6
4.8
4.6

39.5
36.4
33.3

41.2
37.9
34.6

45° C.G.

0° C

94°

A
A

62° 30°–
50° 15° 30°–
50°

B

25.7
23.4
21.8

35.3
32.5
29.6

Ischia

5.6
5.2
4.8

C

Bending
the trunk is
fatiguing.

C

18.1
16.7
15.3

27.9
25.8
23.5

45°

14.9
13.1
11.7

Limit of Color
Discrimination
Visual Limit
(left eye)

21.4
19.7
18

Reach Addition
(arm horizontal)

90°

4.6
4
3.6

Eye Movement:

15° Ea
35° Ma

| Direction | 0° | 45° | 90° |
|---|---|---|---|
| Extend Shoulder A | 2 | 3 | 4 |
| Rotate Trunk B | 0 | 1 | 2 |
| Bend Trunk C | 6 | 8 | 10 |
| Total | 8 | 12 | 16 |

2.9
2.6
2.4

12.6
11.7 Shoe
10.0

Reach Subtraction

17.7
16.2
14.8

| Hand Grip: | 5.4 |
| | 4.8 |
| | 4.4 |
| Finger Grip: | 3.5 |
| | 3.1 |
| | 2.6 |

Standing: 30°
Walking 15°
Striding 7°

4.2
3.9
3.6

4.7
4.4
4.1

3.3
3
2.7

3.7
3.5
3.2

3.7
3.5
3.2

0 Datum

Weight: 192
172 lb (not for health purposes)
151

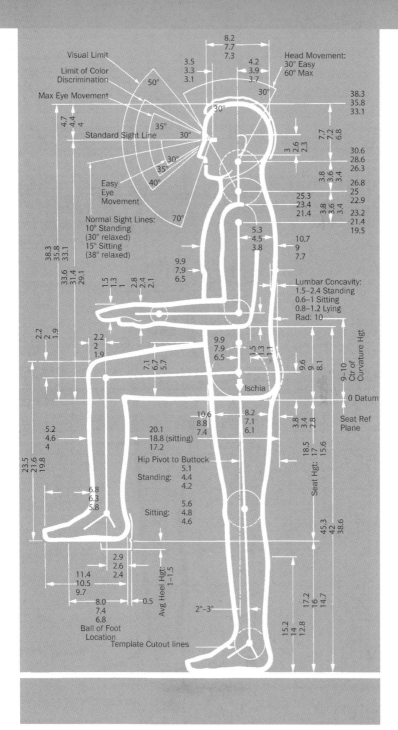

Visual Limit
Limit of Color Discrimination
Max Eye Movement
50°

8.2
7.7
7.3

3.5
3.3
3.1

4.2
3.9
3.7

Head Movement:
30° Easy
60° Max

30°

30°

38.3
35.8
33.1

4.7
4.4
4

35°

Standard Sight Line 30°

7.7
7.2
6.8

3
2.6
2.3

30.6
28.6
26.3

30°

30°

Easy
Eye
Movement

35°

40°

3.8
3.6
3.4

26.8
25
22.9

25.3
23.4
21.4

3.8
3.6
3.4

23.2
21.4
19.5

Normal Sight Lines:
10° Standing
(30° relaxed)
15° Sitting
(38° relaxed)

70°

5.3
4.5
3.8

10.7
9
7.7

38.3
35.8
33.1

33.6
31.4
29.1

9.9
7.9
6.5

Lumbar Concavity:
1.5–2.4 Standing
0.6–1 Sitting
0.8–1.2 Lying
Rad: 10

1.5
1.3
1

2.8
2.4
2.1

2.2
2
1.9

2.2
2
1.9

9.9
7.9
6.5

1.5
1.3
1.1

9.6
9
8.1

9–10
Ctr of
Curvature Hgt

7.1
6.7
5.7

Ischia

0 Datum

10.6
8.8
7.4

8.2
7.1
6.1

3.8
3.4
2.8

Seat Ref
Plane

5.2
4.6
4

20.1
18.8 (sitting)
17.2

Hip Pivot to Buttock

Standing:
5.1
4.4
4.2

18.5
17
15.6

Seat Hgt:

23.5
21.6
19.8

Sitting:
5.6
4.8
4.6

45.3
42
38.6

6.8
6.3
5.8

2.9
2.6
2.4

Avg Heel Hgt:
1–1.5

11.4
10.5
9.7

8.0
7.4
6.8

0.5

2°–3°

17.2
16
14.7

Ball of Foot
Location
Template Cutout lines

15.2
14
12.8

# STATIC AND DYNAMIC MEASUREMENTS

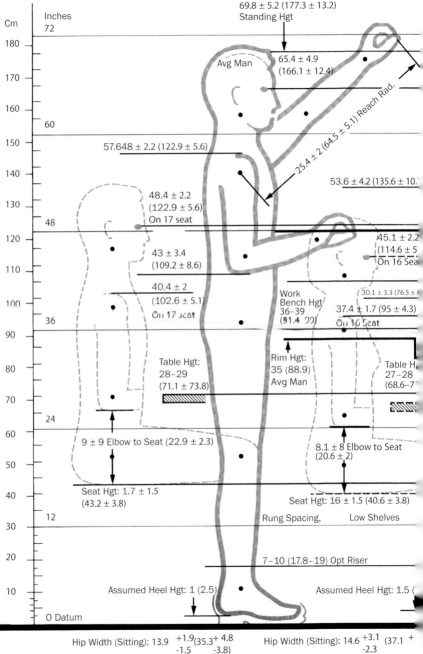

69.8 ± 5.2 (177.3 ± 13.2)
Standing Hgt

Avg Man

65.4 ± 4.9
(166.1 ± 12.4)

25.4 ± 2 (64.5 ± 5.1) Reach Rad.

57.648 ± 2.2 (122.9 ± 5.6)

53.6 ± 4.2 (135.6 ± 10.

48.4 ± 2.2
(122.9 ± 5.6)
On 17 seat

45.1 ± 2.2
(114.6 ± 5
On 16 Sea

43 ± 3.4
(109.2 ± 8.6)

40.4 ± 2
(102.6 ± 5.1)
On 17 seat

Work
Bench Hgt
36–39
(91.4–99)

30.1 ± 3.3 (76.5 ±

37.4 ± 1.7 (95 ± 4.3)
On 16 Seat

Table Hgt:
28–29
(71.1 ± 73.8)

Rim Hgt:
35 (88.9)
Avg Man

Table H,
27–28
(68.6–7

9 ± 9 Elbow to Seat (22.9 ± 2.3)

8.1 ± 8 Elbow to Seat
(20.6 ± 2)

Seat Hgt: 1.7 ± 1.5
(43.2 ± 3.8)

Seat Hgt: 16 ± 1.5 (40.6 ± 3.8)

Rung Spacing,          Low Shelves

7–10 (17.8–19) Opt Riser

Assumed Heel Hgt: 1 (2.5)

Assumed Heel Hgt: 1.5 (

O Datum

Hip Width (Sitting): 13.9 $^{+1.9}_{-1.5}$ (35.3 $^{+4.8}_{-3.8}$)

Shoulder Width: 17.7 + 1.7 (45 + 4.3)

Hip Width (Sitting): 14.6 $^{+3.1}_{-2.3}$ (37.1 +

Shoulder Width: 16 + 1.6 (40.6 + 4.1)

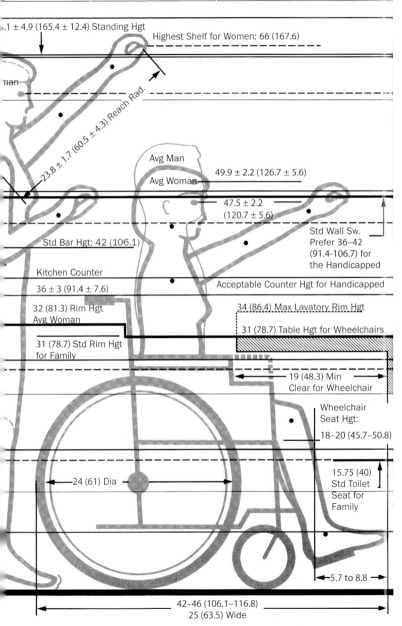

.1 ± 4.9 (165.4 ± 12.4) Standing Hgt

Highest Shelf for Women: 66 (167.6)

nan

23.8 ± 1.7 (60.5 ± 4.3) Reach Rad.

Avg Man

Avg Woman

49.9 ± 2.2 (126.7 ± 5.6)

47.5 ± 2.2
(120.7 ± 5.6)

Std Bar Hgt: 42 (106.1)

Std Wall Sw.
Prefer 36–42
(91.4–106.7) for
the Handicapped

Kitchen Counter
36 ± 3 (91.4 ± 7.6)

Acceptable Counter Hgt for Handicapped

32 (81.3) Rim Hgt
Avg Woman

34 (86.4) Max Lavatory Rim Hgt

31 (78.7) Table Hgt for Wheelchairs

31 (78.7) Std Rim Hgt
for Family

19 (48.3) Min
Clear for Wheelchair

Wheelchair
Seat Hgt:

18–20 (45.7–50.8)

24 (61) Dia

15.75 (40)
Std Toilet
Seat for
Family

5.7 to 8.8

42–46 (106.1–116.8)
25 (63.5) Wide

Designers often reference static dimensions, which refer to measurements taken when the body is at rest (not in motion). Static dimensions are available from a variety of resources. See the anthropometric websites and suggested reading at the end of this chapter. Dynamic dimensions, which reference the body in motion, are also used. However, they are more difficult to measure and are usually captured using mock-ups, measuring, and videotaping. Design issues related to clearance and reach are more appropriately resolved using dynamic anthropometrics. Using real people is always recommended when evaluating mock-ups.

KIDNEY DESK — With regard to reach, a **workspace envelope** is developed to determine how far someone can reach while working at a desk or driving a long-haul truck.

**Body Variations**

**A.** 2.5 percentile female, compared to 97.5 percentile male

**B. Body types:** Thin/Ectomorphic, Muscular/Mesomorphic, and Rotund/ Endomorphic

**C.** Abdominal depth before pregnancy is 6.5" on average, abdominal depth in the last stage averages 11.7"

## C. Pregnancy

## B. Body Types

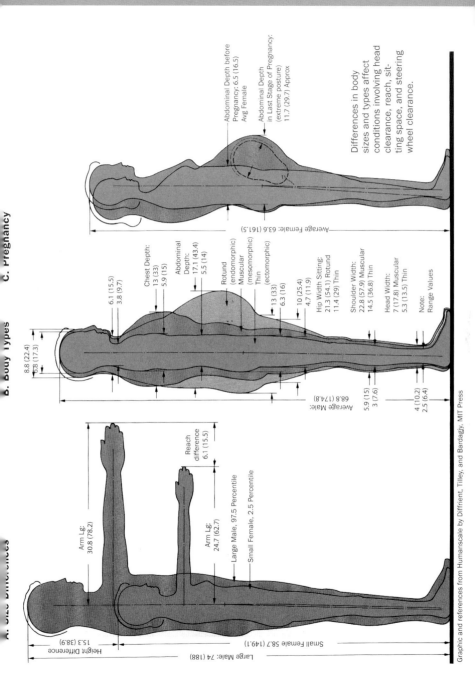

Abdominal Depth before Pregnancy: 6.5 (16.5) Avg Female

Abdominal Depth in Last Stage of Pregnancy: (extreme posture) 11.7 (29.7) Approx

Differences in body sizes and types affect conditions involving head clearance, reach, sitting space, and steering wheel clearance.

Average Female: 63.6 (161.5)

Chest Depth: 13 (33) 5.9 (15)

Abdominal Depth: 17.1 (43.4) 5.5 (14)

Rotund (endomorphic) Muscular (mesomorphic) Thin (ectomorphic)

13 (33) 6.3 (16)

10 (25.4) 4.7 (11.9)

Hip Width Sitting: 21.3 (54.1) Rotund 11.4 (29) Thin

Shoulder Width: 22.8 (57.9) Muscular 14.5 (36.8) Thin

Head Width: 7 (17.8) Muscular 5.3 (13.5) Thin

Note: Range Values

6.1 (15.5) 3.8 (9.7)

8.8 (22.4) 6.8 (17.3)

Average Male: 68.8 (174.8)

5.9 (15) 3 (7.6)

4 (10.2) 2.5 (6.4)

Reach difference 6.1 (15.5)

Arm Lg: 30.8 (78.2)

Arm Lg: 24.7 (62.7)

Large Male, 97.5 Percentile

Small Female, 2.5 Percentile

Height Difference 15.3 (38.9)

Small Female 58.7 (149.1)

Large Male: 74 (188)

Graphic and references from Humanscale by Diffrient, Tilley, and Bardagjy, MIT Press

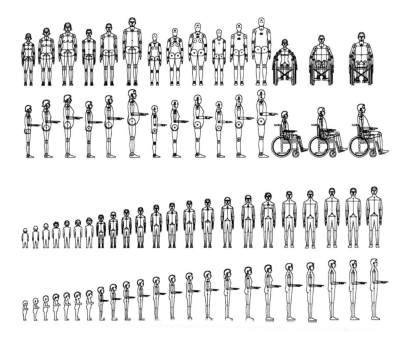

ERGOFORMS V1.0 — Using resources like ErgoForms, designers can insert two-dimensional figures into their CAD programs and develop accurate dimensional relationships.

**Ergonomic and Anthropometric Principles** The following are three general principles for applying anthropometric data.[1]

**Design for the extreme** This accommodates all the population identified.
Take into consideration the largest percentile population (when determining head clearance in an automobile) and the smallest percentile population
(when determining the distance between seat and the pedals of a automobile).

**Design for adjustable range** These designs provide for adjustability, allowing comfortable use for the largest user population. Examples: adjustable work chair or work surface.

**Design for the average** When adjustability is not an option, design for the average height. Example: supermarket checkout counter.

**[1]Source:** Sanders, Mark and Ernest McCormick. Human Factors in Engineering and Design. New York: McGraw-Hill, Inc., 1993, 423.

**Application of Ergonomic Data**[2]

**1.** Determine the body dimensions important in the design.

**2.** Define the population to use the equipment or facilities (children, women, U.S. civilians, different age groups, world populations, etc.).

**3.** Determine what principle in application (extreme, adjustable, or average) should be applied.

**4.** When relevant, select the percentiles of the population to be accommodated, 2.5 percentile, 50 percentile, etc.

**5.** Locate anthropometric tables appropriate for the population and extract relevant values.

**6.** If special clothing is to be worn, add appropriate allowances.

**7.** Build a full-scale mock-up; have people representative of large and small users test and evaluate.

[2]**Source:** Sanders, Mark and Ernest McCormick. Human Factors in Engineering and Design. New York: McGraw-Hill, Inc., 1993, 423.

## LINK EVENT ANALYSIS

Link event analysis is a technique used to determine the best layout for a control panel. It is often used on a piece of equipment where a control panel already exists. But it also has applications as a planning and analysis tool in a variety of settings.

A link is any sequence of use between two points, for example:

• between two control knobs or between a control knob and instructions

• floor patterns or traffic between architectural or interior spaces

• critical path elements for getting a big project done (design, engineering, architecture)

When evaluating links, the value of each link is determined by the frequency of use or importance. The more frequent the use, or importance of control, the higher the value.

The following steps can be followed for a Link Event Analysis Procedure:

**1.** List events: list all the steps a user would take to interact with a product

**2.** Diagram the control panel: draw a diagram of the existing control panel

**3.** Diagram the path followed: draw a diagram of the path the user would follow on the panel

**4.** Diagram linear path followed (2—5—7—1): assign numbers to the controls and show path in a linear format, repeating the numbers as necessary

**5.** Create a linear diagram without repeating steps: redraw this linear diagram, but don't repeat the steps

**6.** Weigh each step: assign each step a weight of 1, if it was repeated once; a 2, if it was repeated twice, a 3, etc.

**7.** Arrange panel in logical order; apply to new concepts.

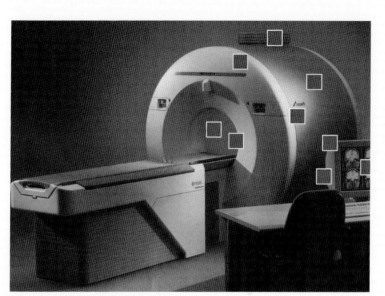

**Link Event Analysis Case Study**

On this Picker Asset MRI Scanner, the following steps were taken to optimize the use of the control panel.

**1.** List events

- Position patient on bed

- Laser light on (positioning light at beginning of tunnel)

- Horizontal drive in (bed moves into tunnel)

- Horizontal drive out (bed moves out of tunnel)

- Tunnel light off (if patient requests it)

- Horizontal drive to scan line (may turn laser light off)

- If hot start is not used technician moves to control room to start imaging

- Horizontal drive out (bed moves out of tunnel)

- Patient moves off bed

continued on next page

continued from page 131

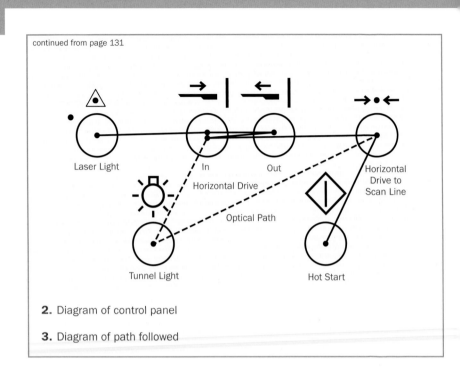

**2.** Diagram of control panel

**3.** Diagram of path followed

**Link Event Analysis Step 4**
**Diagram Linear Path Followed**

1 — 2 — 3 — 2 — 5 — 4 — 6

**Link Event Analysis Step 5**
**Create Linear Diagram Without Repeating Steps**

1 — 2 — 3 — 5 — 4 — 6

**Link Event Analysis Step 6**
**Weigh Each Step**

1 —[1] 2 —[2] 3 — 5 — 4 — 6[1]

**4.** Linear diagram of path followed

**5.** Linear diagram without repeating steps

**6.** Weigh each step

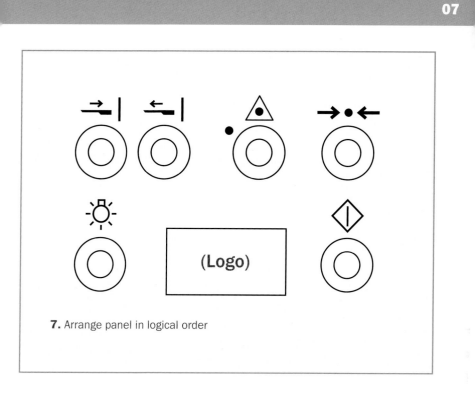

**7.** Arrange panel in logical order

## SAFETY AND HUMAN ERROR

In addition to usability and comfort, other priorities in ergonomics include safety and the prevention of human error and accidents. With the correct data, designers can predict many potential accidents and hopefully minimize the risk in using a product. Placing blame does not motivate identifying human error; rather, the objective is to uncover why an error occurred in the first place, so that corrective action can be taken. Human error is defined as an undesirable human decision or behavior that reduces effectiveness, safety, or system performance.

Designers can predict the potential accidents these cluttered steps might create. Including hand rails at a proper height and keeping them neat minimizes the risk.

When leaving unfinished electrical wiring projects unattended, simple precautions, such as capping wires, turning off power, or adding warning tape, can prevent injuries.

In many cases, accidents are preventable, and design can play a critical role in identifying and addressing potential problems by redesigning flawed products, process, and environments. Three approaches to addressing errors are:

**Exclusion designs** The design makes it impossible to commit the error

**Prevention designs** The design makes it difficult to commit the error

**Fail-safe designs** The design reduces the negative results of the error

## SAFETY LEGISLATION

The Occupational Safety & Health Administration (OSHA) was established in the U.S. in 1970 to create guidelines and enforcement for the safe design of processes, products, and equipment. As a result, working conditions and product safety have improved dramatically.

OSHA tracks accident data relating to human performance, design, and process situations. An example of OSHA findings include data showing that stress conditions, such as heat, fatigue, lack of sleep, consumption of alcohol, noise and toxic gases all contribute to accidents. OSHA regulations can help determine the cause of an accident and predict human behavior in a given situation. Designers can use the available data to minimize the chances of design or situation-induced error.

Specially designed equipment, such as chemical-resistant suits, gloves, and boots, reduce or eliminate risks related to hazardous chemical exposure.

Additional legislation includes the Child Safety Protection Act, which was established in the U.S. in 1995. A major component of the act is a ban on the sale of any ball including marbles or blocks smaller than 1.75 inches, or 44.45 millimeters, for use by children under three years old. Objects under 1.75 inches are considered to be a choking hazard.

A swallow gauge, with dimensions that specifically correlate to the dimensions of a child's throat, was created for testing small parts in children's toys. If the part fits inside the gauge, it is not intended for children under three years. This helps prevent accidental choking.

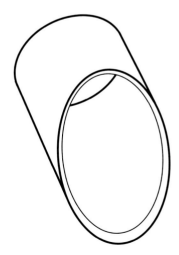

The Child Safety Protection Act also requires the use of a warning label for instances when a toy intended for older children has choking hazards, because many households include older and younger children who may share toys.

## Perceived Risks for Children

These are what parents perceive as the biggest risks for their children.

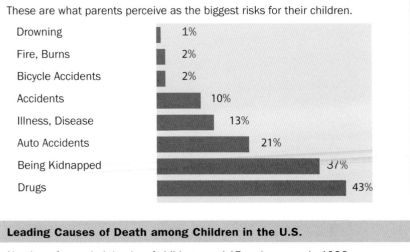

| | |
|---|---|
| Drowning | 1% |
| Fire, Burns | 2% |
| Bicycle Accidents | 2% |
| Accidents | 10% |
| Illness, Disease | 13% |
| Auto Accidents | 21% |
| Being Kidnapped | 37% |
| Drugs | 43% |

## Leading Causes of Death among Children in the U.S.

Number of recorded deaths of children aged 15 and younger in 1986.

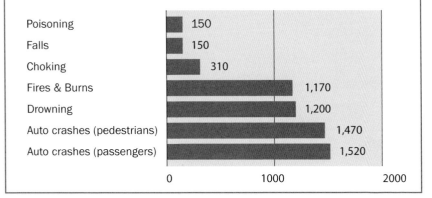

| | |
|---|---|
| Poisoning | 150 |
| Falls | 150 |
| Choking | 310 |
| Fires & Burns | 1,170 |
| Drowning | 1,200 |
| Auto crashes (pedestrians) | 1,470 |
| Auto crashes (passengers) | 1,520 |

0    1000    2000

Interestingly, there are great statistical differences in our evaluation of risk. Actual risks and perceived risks are very different, especially in the likelihood of injury or death. It is essential to differentiate between actual and perceived risks when designing for children.

All toys and games containing small parts that are intended for children ages three to six years old must contain "a clear and noticeable" warning label stating that the toy presents a choking hazard and needs adult supervision. Balloons, whether inflated, uninflated, or broken, are also part of this act and must include instructions that they are not intended for children under 8 years old. Safety laws are almost always the result of tragic circumstances where a product led to the death or injury of young children.

---

**Designing a Reasonably Safe Product[3]**

Products must be designed for reasonably foreseeable use, not solely intended use.

**1.** Delineate the scope of product uses.

**2.** Identify the environments in which the product will be used.

**3.** Describe the user population.

**4.** Postulate all possible hazards, including estimates of probability of occurrence.

**5.** Delineate alternative design features or production techniques.

**6.** Evaluate alternatives relative to the expected performance standards of the product, including:

    **A.** Other hazards that may be introduced by the alternatives

    **B.** Their effect on the subsequent usefulness of the product

    **C.** Their effect on the ultimate cost of the product

    **D.** A comparison to similar products

**7.** Decide which features to include in the final design.

[3]**Source:** Sanders, Mark and Ernest McCormick. Human Factors in Engineering and Design. New York: McGraw-Hill, Inc., 1993, 423.

---

## CHILD DEVELOPMENT

Knowledge of child development will go a long way in helping to design toys that will engage children. Play is a natural activity for every child, and designers can provide many opportunities for growth in different stages. If play is the child's work, then toys are the child's tools. Well-designed toys can help the child work. Keep in mind that young children explore objects in their environment by "mouthing" them. Children can choke to death on such items, especially balloons, balls, and any small household object.

This section offers a look at abilities, interests, play, and pastimes for various age groups.

We can divide infants to the eight-year-old range into separate age groups including:

• Young infants (birth to 6 months)

• Older infants (7 to 12 months)

• Young toddlers (1 year old)

• Older toddlers (2 years old)

• Preschoolers (3, 4, and 5 years old)

• Young children (6, 7, and 8 years old)

**Abilities and interests for young infants (birth to 6 months)**

• Gain control of hands, try to bat, may reach to grasp objects

• Discover feet, bring feet to mouth, may explore with feet

• Begin to sit with support

• Gross motor skills include bouncing, scooting, rocking, and rolling

• Recognize familiar people, objects, and events

• Develop preferences for people, objects, and events

• Smile at faces

• Want attention and contact with people

Interesting toys for this age are for looking, listening, sucking, or touching. Try to include bright primary colors, high-contrast, simple designs, human facial features (especially eyes), and bulls-eye patterns. Infants also enjoy watching hanging objects or mobiles that move by wind or wind-up action.

Toys for watching can be more appealing if they move and make noise. Items for holding should be light and easy to grasp. Mouthable toys should have all safety features recommended for infants:

• no sharp points or edges
• no small parts that can be lodged in throat, ears, or nose
• nontoxic materials
• no long strings to prevent strangling

### Abilities and interests for older infants 7-12 months

• Begin to sit alone, creep, or crawl
• Begin to pull to a stand, cruise along furniture, and walk alone
• Develop pincer (thumb and finger) grasp
• Interest in appearing and disappearing
• Develop object permanence
• Play best with a familiar person nearby
• Sensitive to social approval and disapproval
• Enjoy simple social games: "peek-a-boo," bye-bye"
• Beginning interest in picture books

Consider toys that empty and fill water or sand as well as those that bang, twist, or squeeze. Children of this age also like to stack objects, fit one object into another, and open and close a variety of objects. Construction toys, such as soft rubber blocks and simple puzzles, are good for developing motor skills and perception. The older infant likes different types of balls such as transparent, chime, or action balls that are at least 1 3/4" (4.4 cm) in diameter to prevent choking.

At this stage the baby will lose interest in crib gyms and suspended toys. Be sure to take them away when the child can push up on hands and knees as they become a strangulation hazard.

**Abilities and interests for young toddlers (1 year old)**

- Try to exercise newly acquired physical skills
- Like to climb: can manage small indoor steps
- Combine objects: make simple block structures
- Very curious stage: constant experimentation with objects
- Identify objects by pointing
- Can make marks on paper, scribble spontaneously
- Try to imitate adult tasks
- Expresses affection for others: favorite toys include certain soft toys and dolls

Children at this age like action toys that produce movement or sounds. They are always on the move, running, climbing, and exploring. Push toys with rods are popular. The steady walker enjoys pull toys with a string. Doll carriages or wagons help with imaginative play as well as gross motor skills. Ride-on toys propelled by pushing with feet (no pedals) are a good choice, as well as construction toys with small, lightweight stacking blocks or peg boards.

Dolls or stuffed toys that are soft bodied, lightweight, and easy to hold or grip can help with make-believe play. For creative play the child might enjoy bells or rattles, also large crayons and sturdy paper. Books can include peek-a-boo books with sturdy cloth or cardboard. Touch-me or tactile books help with this learning play.

**Abilities and interests for older toddlers (2 years old)**

- Lots of physical activity such as jumping from heights, climbing, and somersaults
- Throws and retrieves many kinds of objects
- Lots of active play with small objects: explore different qualities of play materials
- Match groups of similar objects
- First counting skills
- First creative activities: process still more important than final product
- Imaginative play increases
- Enjoy hearing simple stories read from picture books, especially stories with repetition; looks at pictures
- Main interest still in parents, but begin cooperative play

Toy choices for the two-year-old can include pull toys with strings, push toys that look like adult equipment (i.e., lawnmower, vacuum, shopping cart), ride-on toys that look like tractors or motorcycles, and toys with storage bins. Tunnels, climbing structures, and slides are ways to develop gross motor skills. Check sports equipment such as sleds; they should be sized to the child (shorter length than child's height.)

Develop fine motor skills and self-help through dressing, lacing, and stringing books and toys. Sand and water play might include bathtub activity centers, water/sand mills, or sprinklers. Puppets representing familiar characters help with creative play, just make sure they are sized to the child's hand. Role-playing and dress-up costumes from Halloween are always popular with two year olds.

### Abilities and interests for preschoolers (3, 4, and 5 years old)

When designing for children ages 3–5, keep mind that they are great doers; they try doing things on their own (like brushing teeth or getting dressed), and they are sympathetic to emotions of others (they offer comfort when others are hurt or sick). At about three years old they can play cooperatively, but gentle play and social interaction is still very difficult for many preschoolers. Children this young still require supervised group play. They love small pets, but are not yet aware of their fragility.

Courtesy of Carla Blackman

Preschool children have won the battle to get on their feet. They are testing the limits of their body. Many are discovering new abilities. Doing something helps them think and thinking helps them do, but it also takes all their concentration. Preschool children are usually pretending to be someone else and pretending to be grown-ups. Props can help out with role-playing, and colorful things are important (they are especially attracted to primary colors). Building, fitting, and counting play are engaging and beneficial developmental milestones. Activities such as counting on an abacus or counting spoons at the dinner table may appear to be a game, but it also has developmental value.

Children continually push their physical limits, often imitating the adult world, which can provide a helpful basis (i.e., mastering a stepladder to reach the sink). Physical skills learned now will seldom be lost, such as swimming, learning to ride a bike, or skating. This is the time they master basic games like hopscotch or skipping.

**Play is important to the 3–5 year old because it:**

- Makes major contributions to physical and emotional growth and development
- Fosters positive self-concept
- Makes sense of the world
- Enhances creativity
- Develops social skills and social thinking

**Abilities and interests of 6-year-old children**

- Children themselves become center of universe, not parents
- Start to think they know everything
- Special playmates are important
- Seek to be first
- Notice leadership
- Marked awareness of gender differences
- Caregivers must use correct vocabulary

The six-year-old is concentrating on favorite play activities. A perpetually popular toy at this age is a bicycle. Children seem to consistently enjoy the locomotor leg exercise and body balance. Both boys and girls enjoy gross-motor activities such as skating, swinging, and running games such as hide and seek, swimming, and ball play. Doll play peaks at age six with much dressing and undressing of dolls and playing with doll accessories, especially for girls.

The six-year-old loves pretending. Playing dress-up helps children imagine themselves as someone else, such as Mom or Dad. Another special interest that continues at this age is transportation toys such as cars, airplanes, trains, and boats. Many children begin collections. They also enjoy painting and coloring, reading simple words, printing letters and numbers, and playing simple board and card games.

**Abilities and interests of 7-year-old children**

• Enter a new phase of growing

• More complaining: some may feel adults don't like them

• Jealousy of siblings becomes more apparent

• Want to prove they are grown up

• Difficult for them to accept blame

• Anxious to please parents, but can end up annoying them

• Need to move growing body

• Like to have a best friend: group play not well organized

Seven-year-old children begin to enjoy playing alone. They love to color and cut, and a craft box is a treasure. Collecting interests continue (i.e., stuffed animals, doll accessories, or action figures).

Coordination is improving in gross motor play; the child becomes a good swimmer, a good batter, and an expert tree climber. In a group the seven-year-old is similar to the six-year-old, but with less ability to pretend and more ability to provide realistic props.

At seven, children love table and board games and jigsaw puzzles. They are less insistent on winning every time and so will take on more complicated games such as Monopoly and magic tricks. The bicycle is still a favorite, and outdoor play is favored over indoors. A seven-year-old is beginning to read more on his/her own but parents should continue to read to him/her whenever they have the opportunity.

## Abilities and interests of 8-year-old children

- Beginning to look mature
- Play is more boisterous: everything is done in high gear
- Wants adult treatment
- Mother still dominant but father becoming increasingly important
- Boys and girls usually play separately
- Reacts to criticism of behavior, but often tries to pass the buck

Active group play, such as sports, clubs, and table games may become the eight-year-old child's favorite play activities. Often, pastimes that are enjoyed with others attract them. They are very organized in their play, for instance organizing a group for a club or arranging elaborate doll play. Boys and girls enjoy all kinds of sports. Playing board and card games with the family is a great opportunity for family bonding and fun. Eight-year-olds also enjoy drama and love acting out a play. Some eight-year-old children enjoy reading and show an interest in popular music.

Toys that interest the eight-year-old are construction sets, doll collections, science kits, arts-and-crafts kits, sports equipment, and board games. They love any kind of collections as they satisfy their developmental need to classify and organize their world.

## What to think about when designing a toy

- Is it safe?
- Does it have small parts that can be swallowed?
- Does it have sharp edges or pinch points?
- Does it suit a child's interests?
- Does the toy challenge without frustrating?
- Does the toy have more than one use?
- Will the toy enable sustained interest?
- Does the toy perpetuate sexist or ethnic stereotypes?
- Does the toy encourage aggressive or violent behavior?
- Does the toy allow the child to participate?

**Developmental areas to keep in mind when designing toys for all ages**

- Gross motor skills (large movements: walking, running)

- Fine motor skills (small movements: building with blocks, etc.)

- Expressive language skills (speech and hearing)

- Receptive communication skills (receiving messages and understanding)

- Social skills (relationships with others: emotional development)

- Self-help skills (leading to independence in "looking after oneself")

- Cognitive adaptive or personal reasoning skills (thinking and involving learning through ones senses)

## Credits:

Irene Bandi. "Characteristics of Elementary School Children." Public School Human Growth and Development Curriculum, Westtlake, Ohio.

Goodson, Barbara and Dr. Martha Bronson. *Which Toy for Which Child*. Washington DC: U.S. Consumer Product Safety Commission, 1993.

Kimberly L. Keith. "Guide to Parenting of K-6 Children." http://childparenting.about.com

## Motor Development Milestones: ages 5 months to 8 years[4]

- **5 months** Rolls over from tummy to back, grasps rattle (Average 25″ long [64 cm], 15 lbs. [33 kg])

- **6 months** Sits briefly without support, stands while holding on, creeps forward

- **7 months** Grasps with thumb and forefinger

- **8 months** Crawls forward

- **11 months** Stands alone, cruises along furniture

- **12 months** Walks alone, can stack blocks (Average 30″ [76 cm] long, 22 lbs [48 kg])

- **13 months** Walks up steps

- **18 months** Speaks 3-50 words, uses feet to scoot along a cycle (Average 31″ [79 cm] long, 23 lbs [51 kg])

- **20 months** Jumps in place

- **24 months** Fits pieces into a puzzle, imitates adults, uses a crayon (Average 32.5″ [83 cm] height, 25.5 lbs [56 kg])

- **2.5–3 years** Jumps a distance of 15 to 24″ (38–61 cm), can ascend stairs unaided, can throw a ball, attempts to write (Average 36.8″ [93 cm] height, 30.8 lbs [68 kg])

- **4 years** Effective control of stopping, starting, and turning, rides a bicycle, learns to swim, can descend stairs unaided (Average 40″ [102 cm] height, 35 lbs [77 kg])

- **5 years** Skips, ties shoes (Average 42.7″ [108 cm] height, 39.4 lbs [87 kg])

- **6 years** Girls are superior in accuracy of movement, boys are superior in forceful, less complex acts (Average 45″ [114 cm] long, 44 lbs [97 kg])

- **7 years** Balancing on one foot without looking, plays hopscotch, jumping jack exercises (Average 47.7″ [121 cm] height, 50.4 lbs [111 kg])

- **8 years** Grip strength of steady 12 lb. pressure, girls can throw small ball 40 ft (12 m) (Average 49.8″ [126 cm] height, 56 lbs [123 kg])

[4]**Source:** Tilley, Alvin R. and Henry Dreyfuss Associates. The Measure of Man and Woman. Hoboken (NJ): John Wiley & Sons, Inc., 2002, pp. 12–21.

The Fisher–Price website, www.fisher-price.com

## TEXT, GRAPHICS, AND SYMBOLS

We depend on our sense of vision to gather 80% of the information around us. Designers need to communicate information through instructions, identifying controls, and signage. That can include written text, graphics, and symbols. We usually encounter various types of text, such as

- **Hard copy**: Text printed on paper
- **Electronic copy**: Digital text on visual display terminals, cell phones, and the like.
- **Signage**: Text applied to large surfaces

The purpose of a sign dictates the specific visual attributes and requirements necessary for the user. However, some criteria to keep in mind for all of them are:

**Visibility**: The quality of a character that makes it separately visible from its surroundings.

**Legibility**: The attribute of alphanumeric characters that makes it possible to identify each one

**Readability**: The quality that makes recognition possible in a grouping

Do:

Don't:

## Type Rules for Products

- Graphic labels for controls typically use uppercase, sans serif fonts (Futura, Helvetica, Univers)
- White letters on a black background are a safe choice, especially in low lighting
- Keep the graphic above the control, so the hand doesn't obstruct the graphic
- Consider important long-term issues

    Will the information wear away over time?

    Will this make the product obsolete?

- Consider what process will be used for applying the graphics

    In-molding

    In-mold decoration

    Pad printing

    Silk-screen printing

    Reverse printed Lexan labels

Courtesy of Daniel F. Cuffaro

This 2005 Mini Cooper interior uses many symbols on its control panel for instant communication across languages.

**Elements to Consider when Designing Graphics for Products**

Take into consideration whether the graphics will be read up-close or at a distance. Design choices are related to stroke width, width to height ratio, size, case (upper, lower, or initial capitals), layout, kerning (interletter spacing), and leading (interline spacing).

**Elements to Consider when Designing Instructions for Products**

Ease of reading, use of symbols, and clear descriptions

- simple sentence construction
- minimal content
- logical order

**Symbols**

Visual symbols and signs are effective in communication without the need for multi-lingual translations. Many international symbols have been developed for the purpose of creating a universal pictorial language. Symbolic signs are typically easier to read/understand and are preferred by viewers.

Recoding—symbols do not need recoding like words do, especially among traffic symbols such as this school crossing sign.

## Objectives of Symbolic Coding Systems

The designer should provide the best possible association of a symbol with its referent.

To establish criteria for selecting coding symbols, do a focus test and ask people the following:

• What does the symbol represent?

• What words match the symbols?

• What are your preferences or opinions?

Look for standardization of symbolic displays; a good source is the IEC Symbols—International Electromechanical Commission (www.iec.ch) or www.graphical-symbols.info/equipment.

AUDIO

AUDIOMICROPHONE

Audio/Microphone
Notice the line weight, drawing style, and negative-positive relationships in these symbols.

## SEATING

The best chairs support the lumbar region, the ischia (sit bones), the thighs, the arms, and the feet. By understanding physical needs, functional needs, purpose, and historical innovations in seating design, it is possible to continue to evolve seating devices to better meet user needs. When there is correct support of the five vertebrae of the spine (known as the lumbar area), it has an inward curve. This is called lordotic. A two-inch thick lumbar support can help achieve this posture. Sitting without a back support causes the back to bend outward in a kyphotic arch, causing increased pressure on the discs, which can lead to damage to that area.

Minimizing pressure on the discs can also be achieved by reclining the seat back by up to 110 degrees. Armrests help relieve pressure on the disks by lifting the weight of the arms off the spine. Sitting in one position for long periods of time, called postural fixity, should also be avoided because a fixed position does not allow the necessary fluid in the discs to be exchanged for spinal health. This same fixity causes static loading of the back and shoulder muscles, along with poor circulation throughout the body. Chairs that allow the user to rock in the chair cause the least amount of stress on the body.

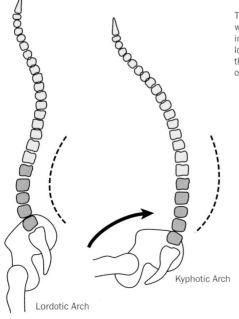

This diagram of the spine with the lumbar region shaded in gray illustrates the correct lordotic posture on the left and the incorrect kyphotic posture on the right.

Kyphotic Arch

Lordotic Arch

## Provide Easy Adjustability

An ideal sitting solution gives many options for adjustability to accommodate as many users as possible. The controls should be easy to reach and clearly marked. They should require no tools and the use of one hand.

## Seat Height

The seat height should be low enough to avoid excessive pressure under the thigh, which can cut off circulation in shorter people. Suggested dimensions include the following:

- 17" (430 mm): Fixed height of the top of the seat pan seat accommodates the largest number of adults

- 13"–17" (330–430 mm): Adjustable height range

Accommodate the fifth percentile of popliteal heights (measurement from floor to inside back of knee when seated); for males, it is 15.5" (395 mm) and for females, it is 14" (355 mm).

## Seat Pan Depth and Width

- 16.8" (427 mm) deep: Maximum depth suitable for small person

- 16"–22" (406–559 mm) wide: Width suitable for large person

A contoured seat pan is also important. Consider that the primary support for the body is on the ischia or sitting bones. To increase the comfort in that area 1"–2" of firm padding is optimal for seats, and gel material can give a custom fit. In this case more is not better. Too much foam can cause the ball and socket joint of the hip to misalign, resulting in discomfort.

The Freedom chair, by Niels Diffrient, is an example of good ergonomic resolution. The arm height, seat height, and angle are easy to adjust and accommodate a variety of users.

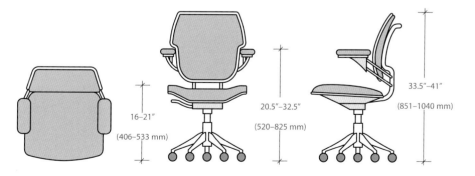

16–21"

(406–533 mm)

20.5"–32.5"

(520–825 mm)

33.5"–41"

(851–1040 mm)

## INGRESS AND EGRESS (ENTER AND EXIT)

How easy is it to get in and out of the seating? It depends on many things including physical strength, flexibility, and energy levels, often referred to as biomechanics. Our center of gravity changes and a freefall effect takes over as we sit down. Take that chair away and what happens? You most likely fall on your bottom. The back legs of a chair need to be angled to prevent tipping when sitting down.

For getting out of a seat, consider elderly individuals, pregnant women, and disabled individuals. The front chair legs should provide space for human feet to support the body while getting out of the chair. Footwear and apparel can also change the dynamics of ingress and egress.

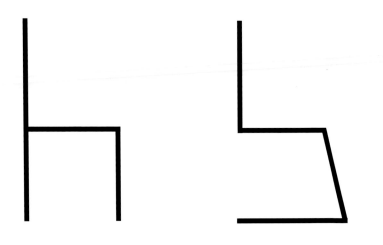

Avoid designing chairs with straight backs and straight back legs. There is a greater tendency for this chair to tip backward and injure the user

Avoid flat or closed fronts. The user must be able to place his/her heels under the lip of the seat in order to comfortably sit or stand.

## COMPUTER WORKSTATIONS

For many of us the computer has become an integral part of daily life. This is a profile of correct posture and positioning of elements around the user. Notice the right angle of the hip and knee joints.

The ideal workstation would have the following:

**1.** An articulated or adjustable Visual Display Terminal (VDT) for flexible options for work height and minimal reflections. An articulating arm lifts the VDT off the work surface for more desk space.

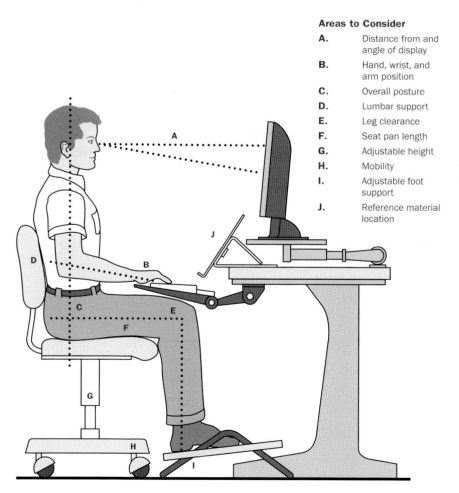

**Areas to Consider**

| | |
|---|---|
| **A.** | Distance from and angle of display |
| **B.** | Hand, wrist, and arm position |
| **C.** | Overall posture |
| **D.** | Lumbar support |
| **E.** | Leg clearance |
| **F.** | Seat pan length |
| **G.** | Adjustable height |
| **H.** | Mobility |
| **I.** | Adjustable foot support |
| **J.** | Reference material location |

**2.** An adjustable tabletop height accommodates a wider range of body sizes. It also allows the user to sit or stand. Good ergonomics gives the body options to move throughout the workday.

Anthro Corporation's Fit System® Adjusta Unit includes a manually adjustable keyboard shelf that encourages the user to adjust his/her positions throughout the day.

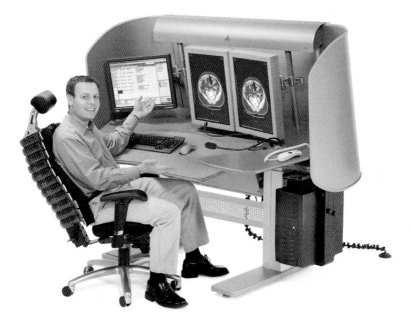

Carl's Table from Anthro Corporation is an electrically adjustable workstation that integrates various factors of comfort including work surface position, lighting, and privacy.

**3.** The tabletop area should be an adequate size for the VDT plus temporary work.

**4.** A distance from eye to screen of 18″ (450 mm) (arm's length) will avoid electro-magnetic fields and minimize eye strain.

**5.** A separate shelf for the keyboard provides adjustable height, moves in and out, and has enough room to accommodate a mouse.

This unit gives the user good adjustability in all areas. The keyboard and mouse should be as close as possible for the least stress on the arm, shoulder, and back.

**6.** A good adjustable chair such as this Freedom Chair by designer Niels Diffrient for Humanscale mimics the spinal column and features an optional Hi Back. Refer to the seating section for further recommendations.

**7.** Arm rests on the chair or separate attachments to the desk help relieve stress on the wrist.

**8.** This footrest angles the foot properly and increases circulation.

**9.** A work holder can reduce eye strain by keeping an equal distance from the eye to paper and eye to screen because the eye doesn't have to refocus as often.

**10.** Good illumination would include task lighting to read hard copy. The ambient lighting should be uniform and neutral—no harsh fluorescents overhead or window reflections in the computer screen.

Courtesy of Carla Blackman

## HANDTOOLS AND DEVICES

The record books are filled with the damage that hand tools can cause. Improper design and usage causes hundreds of thousands of accidents and injuries annually in the U.S. The most common tools to cause injuries are knives, wrenches, and hammers. These usually cause single injury trauma.

Others are known as cumulative trauma, which causes injury from repetitive motion over months or years. It can damage nerves, tendons, ligaments, and joints in the hand, wrist, or elbow. Some common examples are carpal tunnel syndrome, trigger finger, ischemia, and even tennis elbow.

### Environment and Training

Tools cannot be redesigned in isolation. Take a good look at the work environment: Are the heights of the work surfaces correct? Can items be repositioned to avoid injury? Often it is not the tool that is the issue, but the way in which the tool is being used.

### Human Arm and Elbow Anatomy

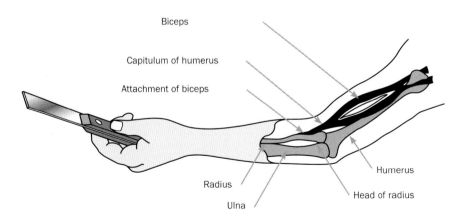

Biceps

Capitulum of humerus

Attachment of biceps

Humerus

Radius

Ulna

Head of radius

# Human Hand Anatomy

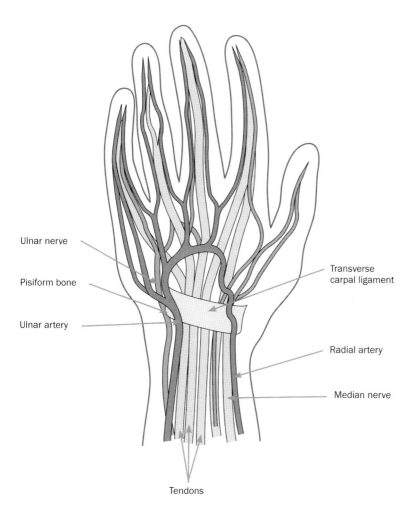

Ulnar nerve

Pisiform bone

Ulnar artery

Transverse
carpal ligament

Radial artery

Median nerve

Tendons

## Wrist Watch—Positions of the Wrist

Ideally, the wrist should be in a neutral position as shown in the diagrams (below). Other positions are called dorsiflexion, palmar flexion, ulnar deviation, and radial deviation. A combination of dorsiflexion and ulnar deviation causes the most damage.

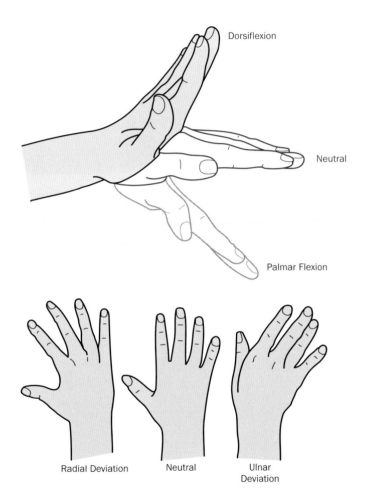

Dorsiflexion

Neutral

Palmar Flexion

Radial Deviation    Neutral    Ulnar Deviation

POSITIONS OF THE WRIST

**Principles of Hand Tool Design**

Maintaining a straight wrist is key in preventing cumulative trauma in the carpal tunnel. The tendons of the fingers pass through the transverse carpal ligament, and they tend to bunch up if the wrist is bent. Carpal tunnel syndrome, another condition of the wrist, causes injury to the median nerve; its symptoms include numbness and loss of strength in the hand.

Bending the handles of the tool helps the hand to stay in a neutral position and can be superior in many applications. Grip strength is reduced if the wrist is bent in any direction. Over several hours this can result in dropping the tool more often or poor workmanship. Grip strength is also related to the size of the object; maximum strength is found in a round object about 1.6″ (40 cm) in diameter.

This shows ulnar deviation in action, putting more strain on the wrist and hand.

## Avoid Tissue Compression

The action of scraping paint off a vertical surface concentrates enough force on the palm to obstruct blood flow to the fingers. This can create numbness and tingling in the hand, forcing the worker to take frequent breaks. To prevent tissue compression in the palm, design handles that have large contact areas or distribute the force over a larger area. Never try to substitute your hand for a hammer; the pounding damage can be considerable to your hands, arms, and even neck. Don't use finger grooves in handles, because they don't fit a large percentage of the population.

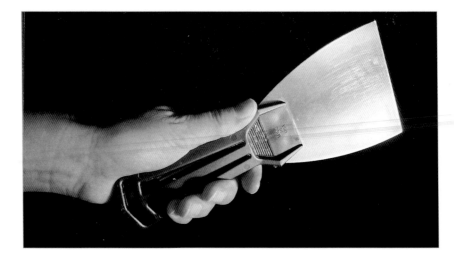

## Avoid Repetitive Finger Action

Overuse of the index finger can result in a condition known as trigger finger. A typical symptom includes the ability to flex but not extend the finger actively. Thumb-operated controls are preferred over finger operated controls.

## Design for Safe Operation

Eliminate pinching hazards and sharp corners or edges for a safer hand tool. Consider adding guards or padding around pinch points and rounding corners. Careful analysis of how the tool will be used and misused is an important part of designing a successful tool.

## Women and Left-handers

In many jobs, such as the military, men and women work alongside each other and are expected to deliver the same results, despite an advantage for males in grip strength and hand size. A good tool design takes into consideration the fact that the average length of a female hand is 80 percent shorter than a male's, and that the average grip strength is 33 percent less than men.

Left-handed people are also at a great disadvantage if a tool has been designed for only right-hand use. Aim to design hand tools that are equally useable in either hand.

## Gloves

Gloves are often used with hand tools for protection against abrasion, cuts, and temperature extremes. They come in a wide variety of sizes, materials, and shapes, and it is important to select the correct glove for the job and the correct fit for hand size. Studies have found that grip strength is often reduced using gloves, which can lead to dropping the tool and additional muscle fatigue.

Many types of gloves can be found on an Ergonomic Safety Supply website, such as www.Conney.com

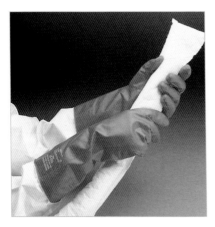

These PVA gloves stand up to strong aromatic and chlorinated solvents that quickly deteriorate rubber, neoprene, and PVC coated gloves. PVA coating is water soluble and not for use in water or water-based products.

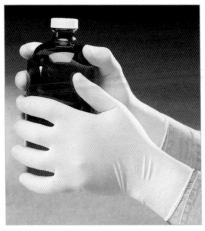

This textured latex glove has fusion bonding that fuses synthetic polymers into the latex, making the glove stronger and more resistant to cracking and peeling. Textured fingers provide a good grip.

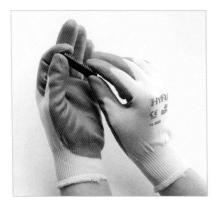

These Hyflex foam-coated gloves are for general assembly and handling small parts for maximum tactile sensitivity. Foam coating is resistant to snags and abrasions, plus it provides excellent grip in light oil applications by channeling oil away from the glove surface. This glove features a knit nylon liner palm coated with a thin layer of nitrile.

## ERGONOMICS AND ANTHROPOMETRIC WEB RESOURCES

**Anthropometric Data of Children**
Ovrt.nist.gov/projects/anthrokids/
child.html

**AnthroTech**
www.anthrotech.net

**Civilian American and European Surface Anthropometry Resource Project—CAESAR**
Store.sae.org

**Cumulative Trauma Disease News**
www.ctdnews.com

**Ergoforms**
www.ergoforms.com

**Ergonomics**
www.ergonomics.com.au

**Human Factors & Ergonomics Society**
www.hfes.org

**NASA Anthropometric Guidelines (based on American males and Japanese females, 2000)**
www.msis.jsc.nasa.gov/sections/sec-tion03.htm

**Occupational Safety & Health Administration**
www.osha.gov

**PeopleSize**
www.openerg.com/psz/index.html

**University of Michigan Series of Ergonomics Links**
www.engin.umich.edu/dept/ioe/C4E/
links.html

**U.S. Military Anthropometric Data Sets**
http://iac.dtic.mil/hsiac/Anthro_US_
Military.htm

**Usernomics**
www.usernomics.com

# Chapter 8: Sustainable Design

by Douglas Paige and Daniel F. Cuffaro

## OVERVIEW

Referencing chapter 1, at its core, industrial design is in the service of business. The historic role of the profession has been to provide a competitive position in the marketplace. However, with the proliferation of products in a growing industrialized world, it has become clear that addressing sustainable issues is critical, and designers have a role and a responsibility to play in the process.

During the last 200 years, humans have created new tools and inventions that have arguably improved the quality of human life, decreased infant mortality rates, and increased standards of living. This in turn has caused a dramatic increase in the global population, resulting in the destruction of habitats, pollution of the environment, and the extinction of many species of plants and animals.

By the late twentieth century, some designers saw that the current path of humanity was unsustainable and began to focus on devising strategies for creating a more sustainable approach to product development. The concept of sustainable design is still in a formative stage but is a growing concern among design professionals. A lynchpin of sustainable design is the concept that the needs of business (profits), the needs of people, and the needs of the planet are not mutually exclusive. This chapter focuses on raising awareness to issues and opportunities related to sustainable design.

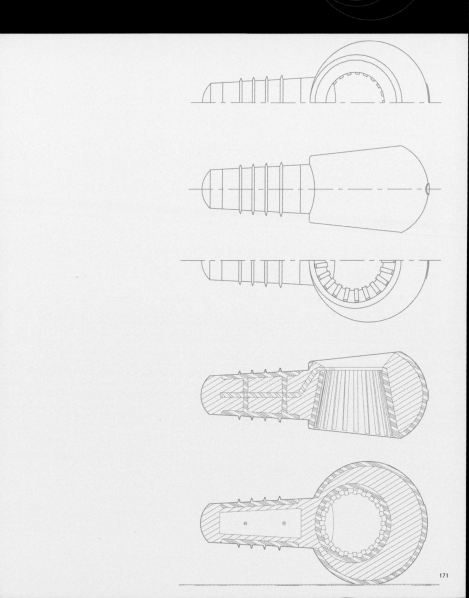

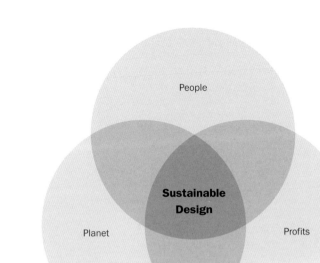

Sustainable design takes into account the health of the planet, people, and company profits.

**The Product Life Cycle** Represents every stage in the creation and use of a product. Understanding the overall cycle dramatically increases the opportunities to address sustainable issues.

1. **Material Selection** What are material options? How has the material been extracted or processed? Is it biodegradable, recyclable, recycled, durable, non-toxic, lightweight, and renewable?

2. **Component Manufacturing** Are environmentally sound practices employed? Are materials and resources used effectively and efficiently? Is production waste minimized or eliminated? Are fair labor practices observed?

3. **Packaging** Is packaging effective and efficient? Are materials reusable or recyclable? Does the packaging inform the user and is it easy to interface with?

4. **Distribution** Have weight and bulk been minimized? Can production happen locally? Is the most effective transportation system being used?

5. **Installation and Use** Is it intuitive and useful? Is it safe? Have the ergonomics been improved? Is it modular or multifunctional? Is it upgradeable and customizable? Are consumables minimized? Does it efficiently use energy resources? Is emission-free or low emission? Is it durable and repairable?

6. **Disposal** Has it been designed for disassembly? Can it be taken back to the manufacturing source (for recycling, reuse, or remanufacturing)? Can the materials be selected for the design of a new product?

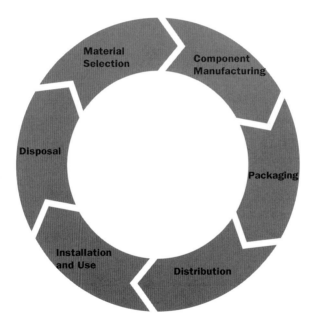

In each stage of the product life cycle, the opportunities to address sustainable issues are numerous, as are the advantages for business. For example, lightweight materials mean lower shipping costs, a healthy workforce is more productive, less packaging lowers costs, useful products build customer loyalty, and reuse of materials lowers raw material cost. Clearly, sustainable design is compatible with profitability. Taking responsibility and effectively planning is necessary for manufacturers, and designers have the ability and responsibility to assume the role of champion for sustainable issues.

## SUSTAINABLE DESIGN STRATEGY

The Ecodesign professional interest section of the Industrial Designers Society of America (IDSA) has published a statement of ecological design principles which can be found at www.idsa.org.

**IDSA recommends the following ecodesign practices:**

### 1. Use ecodesign strategies appropriate to the product

**a.** Reduce overall material content and increase the percentage of recycled material in products.

**b.** Reduce energy consumption of products that consume energy.

**c.** Specify sustainably grown materials when using wood or agricultural materials.

**d.** Design disposable products or products that wear out to be more durable and precious.

**e.** Eliminate unused or unnecessary product features.

**f.** Use materials that are lightweight and therefore use less energy to transport.

**g.** Design for easy, economical disassembly of major components prior to recycling.

**h.** Design products so the that toxic components (electronics et al) can be easily removed prior to recycling.

### 2. Perform comprehensive environmental assessment

**a.** Consider all of the ecological impacts from all of the components in the product over its entire life cycle, including extraction of materials from nature, conversion of materials into products, product use, disposal or recycling and transport between these phases.

**b.** Consider all ecological impacts including global warming, acid rain, smog, habitat damage, human toxicity, water pollution, cancer-causing potential, ozone layer depletion, and resource depletion.

**c.** Strive to reduce the largest ecological impacts.

**d.** Conduct life cycle impact assessment (LCA) to identify opportunities for improving ecological performance.

### 3. Encourage new business models and effective communication

**a.** Support product "take back" systems that enable product upgrading and material recycling.

**b.** Lease the product or sell the service of the product to improve long-term performance and end-of-life product collection.

**c.** Communicate the sound business value of being ecologically responsible to clients and commissioners.

**d.** Discuss market opportunities for meeting basic needs and reducing consumption.

**e.** Present superior product quality claims (energy saving, contains less toxic waste, etc.) along with other performance features.

## RESOURCES

**Centre for Environmental Assessment of Product and Material Systems (CPM)** www.cpm.chalmers.se

**Centre for Sustainable Design**
www.cfsd.org.uk

**Design for the Environment Multimedia Implementation Project (DEMI)**
www.demi.org.uk

**Ecodesign Foundation**
www.edf.edu.au

**European Design Centre**
www.edc.nl

**Industrial Design Society of America (IDSA)**
www.idsa.org

**McDonough Braungart Design Chemistry**
www.mbdc.com

**02 Network**
www.02.org

# Chapter 9: Design Documentation

### by Douglas Paige

## OVERVIEW

Communication of an idea or a specific design is critical to the product development process. Designers communicate with each other, end users, marketers, engineers, and manufacturers using many different visual tools, including sketches, renderings, models, design control drawings, and digital files. A detailed hand drawing or computer-generated control drawing is critical in the communication of design intent for model making, engineering, and manufacturing. Effective control drawings include accurate part or product views, sections, and dimensions.

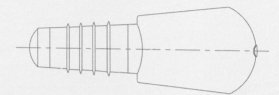

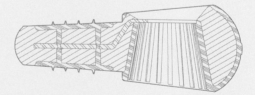

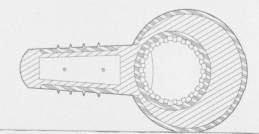

## ORTHOGRAPHIC DRAWINGS

Standard Formats and Sizes

In the U.S., there are two standard drafting formats: ANSI (American National Standards Institute) for engineering drawings, and ARCH for architectural drawings. Each size has a letter designation, A through E, and each successive page is double in size. These dimensions represent the paper size, while the borders are drawn 3/8″ (9 mm) or 1/2″ (13 mm) from the edge of the paper. In addition, the metric standard, ISO (International Organization for Standardization), is the common international standard.

### U.S. Engineering

| Size | inches | mm equivalent |
|------|--------|---------------|
| ANSI A | 8.5 x 11 | 216 x 279 |
| ANSI B | 11 x 17 | 279 x 432 |
| ANSI C | 17 x 22 | 432 x 559 |
| ANSI D | 22 x 34 | 559 x 864 |
| ANSI E | 34 x 44 | 864 x 1118 |

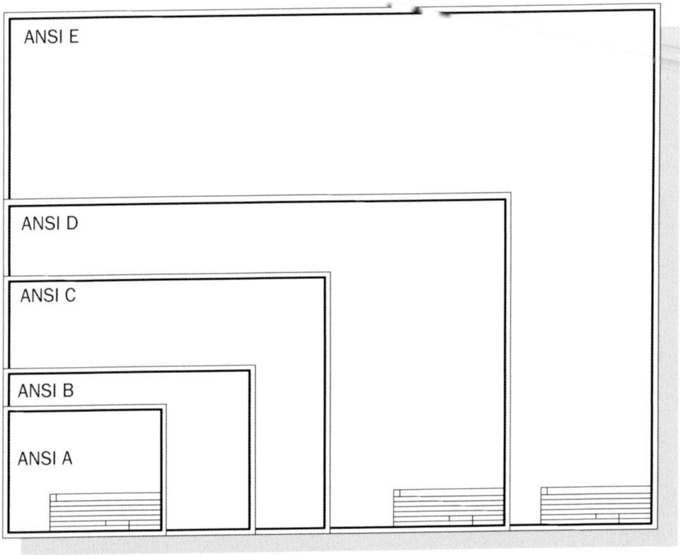

ANSI E

ANSI D

ANSI C

ANSI B

ANSI A

## U.S. Architectural

| size | inches | mm equivalent |
| --- | --- | --- |
| ARCH A | 9 x 12 | 229 x 305 |
| ARCH B | 12 x 18 | 305 x 457 |
| ARCH C | 18 x 24 | 457 x 610 |
| ARCH D | 24 x 36 | 610 x 914 |
| ARCH E | 36 x 48 | 914 x 1219 |

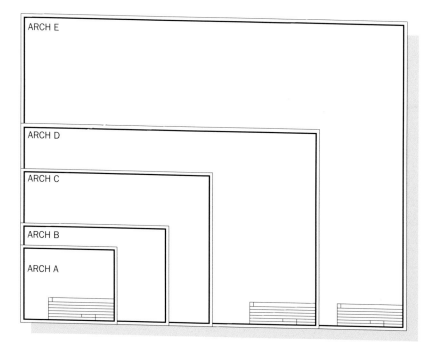

### International Standard

| ISO | (mm) | inches equivalent |
|-----|------|-------------------|
| A5 | 148 X 210 | 5.8 x 8.3 |
| A4 | 210 X 297 | 8.3 x 11.7 |
| A3 | 297 X 420 | 11.7 x 16.5 |
| A2 | 420 X 594 | 16.5 x 23.4 |
| A1 | 594 X 841 | 23.4 x 33.1 |
| A0 | 841 x 1189 | 33.1 x 46.8 |

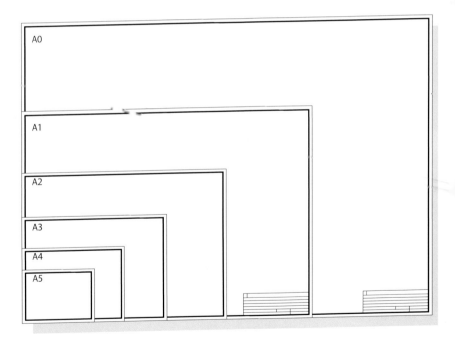

## LAYOUT

An orthographic drawing shows an image of an object from several views on one page. Each view (surface) is drawn as it would appear looking straight at it, with no perspective or other distortion to the image. The primary view for the object is generally referred to as the front view. All other views are located horizontally or vertically in line with the front view. Each adjacent view shows what the object would look like as you rotate it 90 degrees.

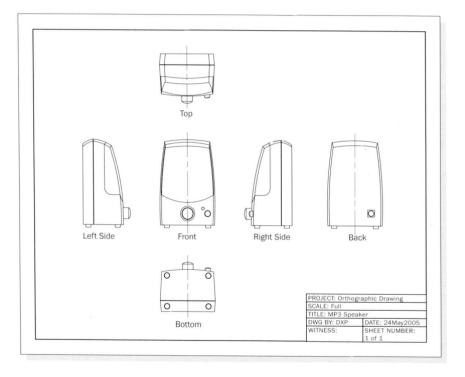

A standard orthographic layout contains a horizontal and vertical axis with the primary (front) view at the intersection. Side and back views are located along the horizontal axis, while the top and bottom views are located on the vertical axis. A common European practice has been to use another layout called first surface projecting. This reverses the position of the left and right views as well as the top and bottom views in relation to the front view.

One way to think about the orientation of the views on an ANSI-compliant orthographic drawing is to consider the object as a box that has been opened up and unfolded.

## Line Weights and Styles

Orthographic drawings use a variety of line weights and styles. Styles are used to signify what the line is describing, while line weight is used to help describe each style and illustrate differences with the solid line.

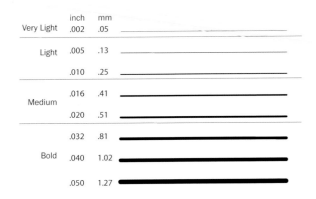

| | inch | mm | |
|---|---|---|---|
| Very Light | .002 | .05 | |
| Light | .005 | .13 | |
| | .010 | .25 | |
| Medium | .016 | .41 | |
| | .020 | .51 | |
| | .032 | .81 | |
| Bold | .040 | 1.02 | |
| | .050 | 1.27 | |

Line weights are measured in simple, subjective terms when drawn by hand: very light, light, medium, and bold. With the use of computers, line weights are now measured in many thicknesses, usually in a dimensional scale:

.05–1.27 mm or .002–.050"

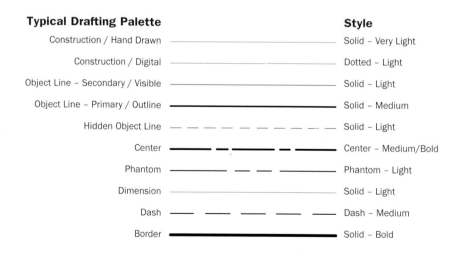

## Type

Construction / Hand Drawn —————————————————

Construction / Digital —————————————————————

Solid —————————————————————

Hidden Object — — — — — — — — — —

Center ——— — ——— — ———

Phantom ————— — — —————

Dash —— —— —— —— ——

Various line types are used to convey different kinds of information in a control drawing. Solid lines can be used for many purposes depending on the weight, including construction, object, dimensions, borders, and notes. Other types have dedicated purposes and are commonly used in single line weights.

## Typical Drafting Palette

| | | Style |
|---|---|---|
| Construction / Hand Drawn | ————————————— | Solid – Very Light |
| Construction / Digital | ————————————— | Dotted – Light |
| Object Line – Secondary / Visible | ————————————— | Solid – Light |
| Object Line – Primary / Outline | ————————————— | Solid – Medium |
| Hidden Object Line | — — — — — — — — — — | Solid – Light |
| Center | ——— — ——— — ——— | Center – Medium/Bold |
| Phantom | ————— — — ————— | Phantom – Light |
| Dimension | ————————————— | Solid – Light |
| Dash | —— —— —— —— —— | Dash – Medium |
| Border | ————————————— | Solid – Bold |

The combination of line type and weight make up the style. This chart depicts a typical palette of line styles used in a design control drawing.

**Construction (very light solid lines or light dotted lines)** Used for the initial layout and as reference points within a drawing. Construction lines are for reference only and are not intended as visible object lines in the final drawing.

**Object (solid lines, either light or medium weight)** Describe the actual physical part. Any visible hard edge in a particular view should be represented by an object line. This includes surface changes, however subtle, so long as there is a hard edge. The primary lines, including major form breaks within the object and the outline edge of the object, use a medium weight line, while secondary surface breaks and openings for buttons or other details use a lightweight object line.

**Phantom (light phantom line)** A surface that changes with a fully tangent radius does not appear as a hard edge but can be defined as a phantom line. These are located along the tangent points where a radius meets an adjacent surface.

**Hidden object (light hidden lines)** Used to describe an edge or surface break in the object that is behind another surface in a particular view. Hidden lines are primarily used for details that are visible object lines in other views with the drawing.

**Center (medium or bold center lines)** Used along the primary x- or y- axis in the center of the part. Center lines are used as reference points from which to create other lines or in dimensioning. Center lines can also be used to describe where a section line cuts through a part.

**Dimension (light solid lines)** Drawn from the part out to the space between views to indicate distance or location of specific elements of a drawing.

**Dash (medium dash lines)** Used for outlining a detail view.

**Border (bold solid lines)** Used around the title box and perimeter of the drawing.

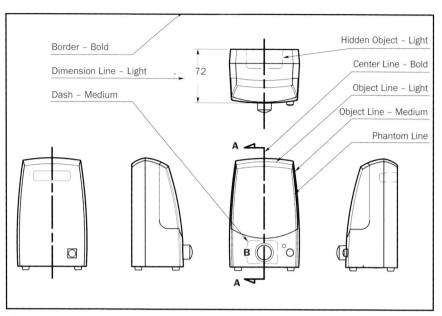

The image above illustrates some examples of line styles used in a design control drawing.

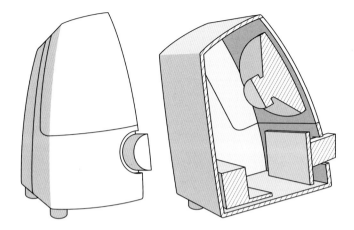

When creating a section view, imagine taking the finished product and cutting it in half. The appropriate visible elements within the open part should be included in the view.

## Section Views

Section views are drawn as if part of the drawing has been cut away and you are looking into the remaining object. These auxiliary views are helpful in explaining details such as recessed surfaces, openings, spacing of parts, and internal details.

## Section Lines

The surface that is "cut" through with the cutting plane becomes the front surface in the section view. This surface should be marked with section lines: thin solid lines at 45 degrees on 1/8″ (3 mm) spacing. This was common practice with hand drawings. However, it has largely disappeared with digitally generated drawings.

## Detail Views

When details are small or too much information is contained in a small area, it becomes necessary to enlarge a portion of the drawing and show this off to one side of either the horizontal or vertical axis. These details are usually enlarged by a factor of two or four depending on the detail. Detail views are outlined in one of the primary views by a dashed line and given a letter designation as with the section views. See Detail B in the diagram (opposite, above).

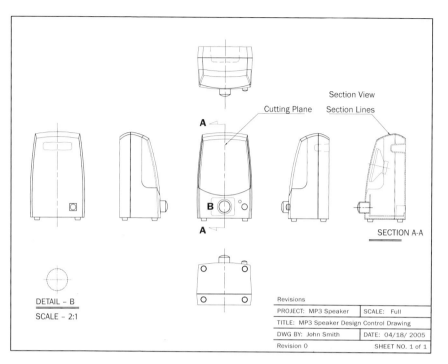

DETAIL – B
SCALE – 2:1

| Revisions | | |
|---|---|---|
| PROJECT: MP3 Speaker | | SCALE: Full |
| TITLE: MP3 Speaker Design Control Drawing | | |
| DWG BY: John Smith | | DATE: 04/18/ 2005 |
| Revision 0 | | SHEET NO. 1 of 1 |

Cutting Plane — Section View — Section Lines

A

B

A

SECTION A-A

A section view is shown by a heavy center line indicating where the part is "cut." Solid lines and arrows indicate the direction of the view. In the example shown, the view is from the right side, so the section view is located past the right side view in the x-axis. Section views are designated by capital letters (A) at each arrow. The description on the corresponding view is shown as "Section A-A," and each subsequent section, or detail view, is indicated by the next letters in the alphabet.

## Projected Views

In some instances, the straight horizontal or vertical axis does not clearly describe the form or details relative to an angled surface. In these cases an auxiliary view must be drawn "projected" perpendicular to the view in question. This will create a new view looking straight at the surface of the object. This is also referred to as "multi views without rotation." The key to an orthographic drawing is the right angle (90 degrees) projection. These auxiliary projected views adhere to that principal of defining views.

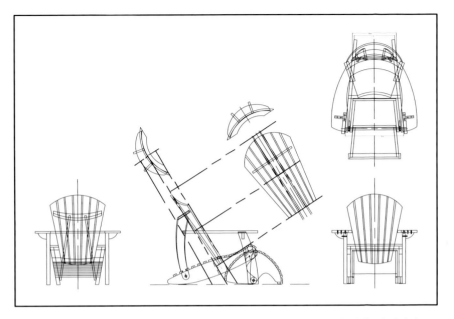

The projected view shows a clear definition of the seat back and upper support in this Adirondack chair.

## Dimensioning

Before the advent of computers, dimensioning was a critical component of a design control drawing. The drawing was used as a visual reference, but the dimensions were the controlling factors for engineers and model makers. Every surface and detail required a dimension to define it clearly. Detailed dimensions are not common with current technology, but there will be cases where a dimensioned drawing is required. Further, there is often the need to put overall dimensions or key part dimensions on a drawing for review purposes. It is important to note that an accurate control drawing will make the dimensioning step much easier. This is still true with a computer file drawing and in cases where a file is being transferred to engineering without dimensions. The engineer is still opening the file and taking dimensions directly from the data. The following is a simple summary of basic dimensioning techniques and styles.

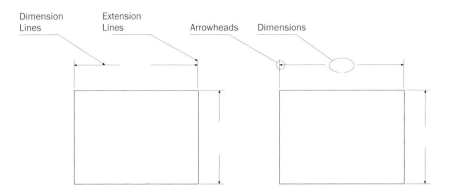

Four basic elements make up most dimensions: dimension lines, extension lines, arrowheads, and numerical measurements. All dimension lines are lightweight solid lines and should not interfere visually with the object in the drawing. Lines should be lighter or thinner than the secondary object lines.

**Dimension lines** indicate direction and extent with the ends terminated by arrowheads at the extension lines. The dimension is placed either in line with the dimension line or horizontally.

**Extension lines** extend from the dimension line to the point on the object in question. A small gap of 1/16″–1/8″ (1.5–3.2 mm) is left between the reference point and the end of the extension line. The outer end extends past the dimension line arrowhead about 1/8″ (3.2 mm).

**Arrowheads** mark the extent of the dimension lines, at intersection with extension lines.

**Dimensions** are located either above or in a break in the dimension line.

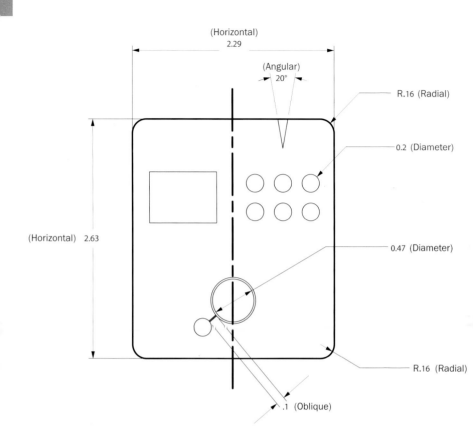

Various types of dimensions can be used, including horizontal, vertical, oblique, radial, diametral, and angular. Note: Extension lines may cross the edge of the object to indicate a measurement inside the part, but the dimension line and dimension shall always be outside of the object. Arrowheads will be outside the object on horizontal and vertical dimensions but could be inside the object for some radial or diameter dimensions.

## Dimensioning Techniques

**Baseline dimensioning** follows the basic example with dimension lines and extension lines. Multiple dimensions are indicated from a common starting point or baseline. The dimensions share one common extension line on the baseline, with individual dimension lines. Dimension text can be placed *in line* or *horizontal*.

**Chain dimensioning** indicates a series of adjacent dimensions, horizontally or
vertically. Adjacent dimensions share a common extension line. Dimension text can
be placed *in line* or *horizontal*.

**Ordinate dimensioning,** also called *datum dimensioning, is* a different method using
a base or datum point marked by a single extension line with a 0 at the end. The 0
datum is commonly one edge of the part or a centerline. Additional dimensions are
indicated relative to the 0 datum by individual extension lines with the dimension
at the end. Only one extension line per dimension and no dimension lines are used
with this technique, leaving a simpler look to the overall dimensioning. Ordinate
dimensioning text works best *in line*.

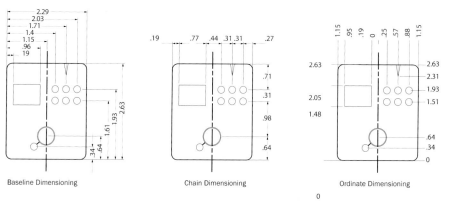

The same drawing is dimensioned with 3 different techniques to illustrate how different techniques
can look depending on the spacing of the details. The dimension text can be placed *in line* or
*horizontal* with *baseline* and *chain* dimensioning. *Ordinate* dimensioning text works best in line.

**Arrowhead styles** At one time arrowheads followed a very consistent pattern. Arrowheads were the height of the dimensions, and the ratio of length to height was 3:1. With 1/8″ (3 mm) being a common height for dimension text, that made a full arrowhead 3/8″ x 1/8″ (9.5 x 3.2 mm). The biggest variable was solid vs. open and full arrowhead vs. half arrowhead. The constant element was the ratio of width to height.

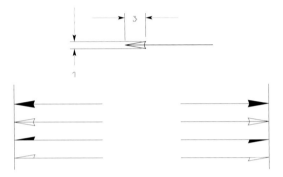

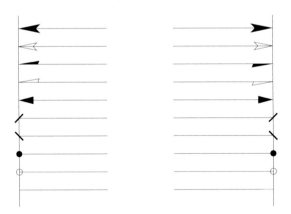

Full arrowheads and half arrowheads in solid or outline style. The arrowhead is reversed on each side of the half arrowhead style.

Current drafting arrowhead styles include variations that are more contemporary. Most of these accomplish the same thing, and this is largely up to the individual preference of the designer and company standards.

## DIMENSIONING

Dimensions and tolerances are important for the correct functioning of the part. There are standards that are used to correctly dimension a drawing. A logical approach to dimensioning is based on an understanding of how the parts will be produced, (e.g., by milling machine, lathe etc.) All dimensions need to be on the drawing along with all other information that is necessary to completely define the part in its final form.

Dimension line is a light full line used to indicate the measurement the amount of which is denoted by figures in a space left in the dimension line or above the dimension line.

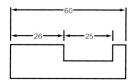

Extension lines (projection lines) are light full lines extending perpendicular to outline and are used when placing dimensions outside the object. There should be a gap of about 2 mm outside the outline and the line should extend about 3 mm beyond the dimensioning line. They could be oblique where necessary but preferably parallel to each other.

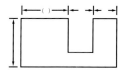

Arrow-heads are used to terminate dimension lines. Its length is about three times the depth and the space should be filled in. Adjacent arrow-heads may be replaced by clearly marked dots where space is limited.

For dimensioning in limited space, the arrow-heads should be reversed.

When there are several parallel dimension lines the figures should be staggered to avoid confusion.

Leaders (pointer lines) are lines drawn from notes and figures to show where these apply. These are thin straight lines terminated by arrow-heads or dots. Arrow-heads should always terminate on a line; dots should be within the outline of the object. The leaders should be curved or made free hand. The other end may terminate in a short horizontal bar at the mid-height of the lettering of the first or last line of the note.

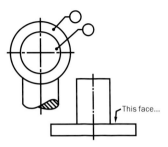

This face...

The two recommended systems of placing the dimensions are: (a) Aligned system (b) Unidirectional system.

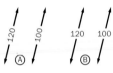

Wherever possible, the dimensions of small radii should be shown outside the outline of the object and should be followed by the letter *R*.

Where the center point of a radius is dimensioned, a dot should be placed at the intersection of the lines which locate the centre of the radius.

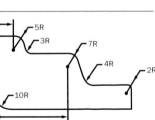

When space is restricted the method shown here can be used to place the dimensions of different diameters.

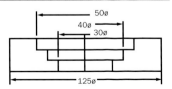

Where it is not possible to show the center of an arc conveniently in its correct position, the method illustrated here should be used. The portion of the dimension line which touches the arc should be in line with the true center.

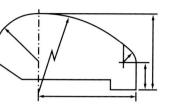

The circles should be dimensioned by one of the methods illustrated here.

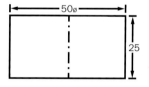

Where it is not obvious that the dimension represents a diameter or where the dimension is remote from the circle, the use of the symbol ø is very essential.

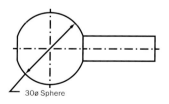

The diameter of a spherical surface should be dimensioned as shown here.

30ø Sphere

## Taper, Keyways and Chamfers

The following dimensions may be used in suitable combinations to define the size and form of tapered features: (a) Diameter (or width) at each end. (b) Length. (c) Diameter (or width) at a selected cross-sectional plane which may be within or outside. (d) Dimension locating a cross-sectional plane at which a diameter or width is specified. (e) The rate of taper or the included angle.

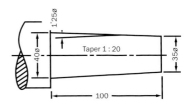

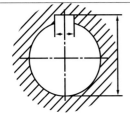

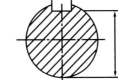

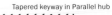

Parallel hub

Parallel shaft

Tapered keyway in Parallel hub

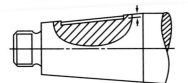

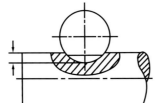

Parallel keyway in tapered shaft

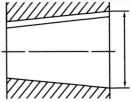

Parallel keyway in tapered hub

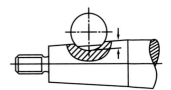

Woodruff keyway in parallel shaft

Woodruff keyway in tapered shaft

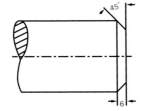

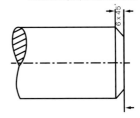

Dimensioning chamfers

**Dimensioning to an intersection with a radius edge** One problem with simply dimensioning to endpoints with computer files is that the dimension of a surface containing a radius at one or both ends will not be accurate in most cases. The dimension should go to the corner or "intersection" before the radius is entered. To get an accurate dimension, you need to extend the primary surfaces to their inter-section point. The part can be trimmed after dimensioning, or the extended lines can be converted to construction lines.

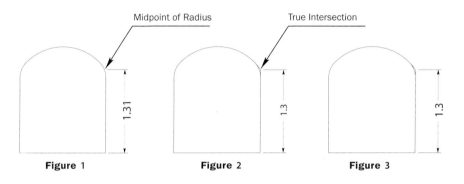

In figure 1, dimension lines will snap to endpoints or midpoints of the radius resulting in an inaccu-rate dimension. Figure 2 represents correct dimensioning. Figure 3 shows extended lines that have been converted to less visible construction lines.

**Fractions and decimals** Fractions were once common with the English measure-ment system. However, they have been largely replaced with decimal equivalents in most design control drawings. With the global nature of manufacturing as it is today, many companies are also using the metric system to allow for easier transitions from design to manufacturing. It is highly recommended to become familiar with the met-ric system for design control drawings. Some industries are not changing as fast as others, while much of the consumer electronics industry has been metric for some time. Please refer to chapter 11 for conversion charts.

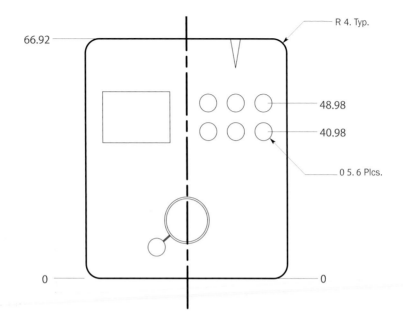

Text can be added to the dimension to add information. For example, a corner radius can be followed with typ., indicating that the radius is "typical" throughout the drawing unless otherwise noted. A diameter dimension can be followed with X plcs., indicating that the same diameter applies to group of identical circles in a drawing.

**Scale drawings** As stated earlier in this chapter, drawings should be full scale (1:1) whenever possible. When the part is too large for any of the standard drawing formats, the part size should be reduced by common factors (1:2, 1:4, 1:5, 1:10, etc.). When objects are very small, it may be difficult to discern details. In this case, it may be necessary to draw the object 1:1 with an additional enlarged view (2:1, 4:1, etc.) to help explain the object more clearly. When drawn by hand, these adjustments can be made with the use of an "engineers' scale", which has various scale measurements on each of 6 surfaces.

When created in a computer file, drawings are usually drawn full scale and reduced to fit a page for plotting out when necessary. It is important to keep in mind that all dimensions need to be the actual 1:1 or full scale of the actual part. It is also important to note the scale in the title box of the drawing.

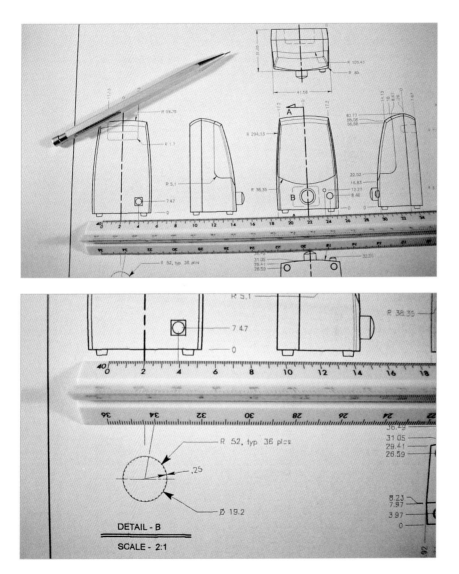

Detail B (above) has been drawn at a scale of 2:1 to show edge detail more clearly. The engineering scale allows the designer to draw and measure scaled details directly, without the need of mathematical conversions.

**Title box** A box is created in the lower right corner of each drawing listing important information for the drawing, including drawing title, project, designer name, date, scale, and revisions. Some companies use very extensive title boxes for enhanced internal tracking information, but these can vary widely by company. The basics are included for this example. The section for revisions can be added above the main title box or to the side. More revisions can be added as needed during the project.

| Revisions | |
|---|---|
| **PROJECT:** MP3 Speaker | **SCALE:** Full |
| **TITLE:** MP3 Speaker Design Control Drawing | |
| **DWG BY:** John Smith | **DATE:** 04 / 18 / 2005 |
| **Revision** 0 | **SHEET NO.** 1 of 1 |

## CALCULATING VOLUMES

Volume calculations are important to product designers. For example, you may have a project to design a coffee maker. To start the project and understand the proportions, you will need to realize the dimensions of the carafe and the water container within the device. Basic calculations of the volume will give you an idea of the overall dimensions of the glass carafe. Yes, when you turn to your CAD software in a more advanced stage of the design, the software will generate the volume. But at the earlier design stage, this basic knowledge ensures that you stay within the parameters of your design brief.

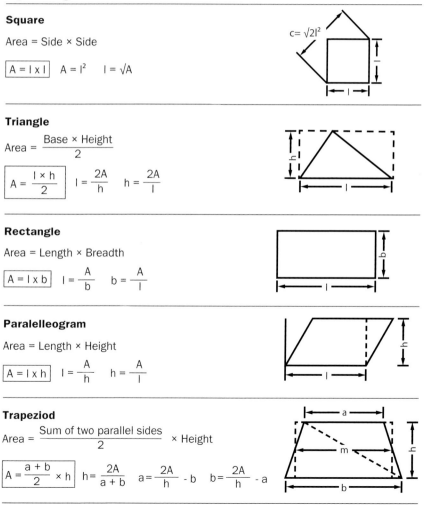

### Square

Area = Side × Side

$$\boxed{A = l \times l} \quad A = l^2 \quad l = \sqrt{A}$$

$$c = \sqrt{2l^2}$$

### Triangle

Area = $\dfrac{Base \times Height}{2}$

$$\boxed{A = \dfrac{l \times h}{2}} \quad l = \dfrac{2A}{h} \quad h = \dfrac{2A}{l}$$

### Rectangle

Area = Length × Breadth

$$\boxed{A = l \times b} \quad l = \dfrac{A}{b} \quad b = \dfrac{A}{l}$$

### Paralelleogram

Area = Length × Height

$$\boxed{A = l \times h} \quad l = \dfrac{A}{h} \quad h = \dfrac{A}{l}$$

### Trapeziod

Area = $\dfrac{Sum\ of\ two\ parallel\ sides}{2} \times Height$

$$\boxed{A = \dfrac{a + b}{2} \times h} \quad h = \dfrac{2A}{a + b} \quad a = \dfrac{2A}{h} - b \quad b = \dfrac{2A}{h} - a$$

## Regular Polygon

Area = Area of each triangle × No. of sides

$$A = \frac{l \times h}{2} \times n$$

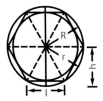

### Calculation of regular polygons

The area can be calculated from l or R (radius of circumscribed circle) or r (radius of inscribed circle)
The length of a side l can be calculated from R or r
The radius R can be calculated from l or r, the radius r from R or l

| n = no. of sides or = no. of corners | Area A = | | | Side 1 = | | Radius R of circumscribed circle | | Radius r of the inscribed circle | |
|---|---|---|---|---|---|---|---|---|---|
| | l² times | R² times | r² times | R times | r times | l times | r times | R times | l times |
| 3 Sides | 0.4330 | 1.2990 | 5.1962 | 1.7321 | 3.4641 | 0.5774 | 2.0000 | 0.5000 | 0.2887 |
| 4 Sides | 1.0000 | 2.0000 | 4.0000 | 1.4142 | 2.0000 | 0.7071 | 1.4142 | 0.7071 | 0.5000 |
| 5 Sides | 1.7205 | 2.3776 | 3.6327 | 1.1756 | 1.4531 | 0.8507 | 1.2361 | 0.8090 | 0.6882 |
| 6 Sides | 2.5981 | 2.5981 | 3.4641 | 1.0000 | 1.1547 | 1.0000 | 1.1547 | 0.8660 | 0.8660 |
| 8 Sides | 4.8284 | 2.8284 | 3.3137 | 0.7654 | 0.8284 | 1.3066 | 1.0824 | 0.9239 | 1.2071 |
| 10 Sides | 7.6942 | 2.9389 | 3.2492 | 0.6180 | 0.6498 | 1.6180 | 1.0515 | 0.9511 | 1.5388 |
| 12 Sides | 11.1960 | 3.0000 | 3.2154 | 0.5176 | 0.5359 | 1.9319 | 1.0353 | 0.9659 | 1.8660 |

---

## The Pythagoras Theorem

The square of the hypotenuse is equal to the sum of the squares of the other two sides.

c = hypotenuse—opposite side to right angle
a, b = two sides forming right angle

$$c^2 = a^2 + b^2 \qquad c = \sqrt{a^2 + b^2} \qquad a = \sqrt{c^2 - b^2} \qquad b = \sqrt{c^2 - a^2}$$

Example:    a = 6cm        b = 8cm        c = ?

$$c = \sqrt{a^2 + b^2} \quad = \sqrt{36 + b^2} \quad = \sqrt{100} = 10cm$$

## Circle

$$A = \frac{\pi}{4} D^2$$   Area $= \frac{\pi}{4} \times$ Diameter $\times$ Diameter

$$\boxed{A = \pi r^2}$$   Circumference $= C = \pi \times$ Diameter
$$\boxed{C = \pi D}$$   $= 3.142D$
$= 2xr$

## Hollow circle

A = Area of greater circle – Area of smaller circle

$$= \frac{x}{4} D^2 - \frac{x}{4} d^2$$

$$\boxed{A = \frac{\pi}{4} (D^2 - d^2)}$$

$$= 0.785 (D^2 - d^2)$$

## Sector of circle

$$A = \frac{\text{Arc length} \times \text{Radius}}{2}$$

$$\boxed{A = \frac{\pi \times r^2 \times \theta}{360}}$$

$$\boxed{A = \frac{\pi}{4} \times \frac{D^2 \times \theta}{360}}$$

$$\boxed{b = \frac{\pi \times D \times \theta}{360}} \qquad \boxed{A = \frac{b \times r}{2}}$$

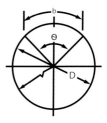

## Segment of circle

A = Area of sector of circle – Area of triangle

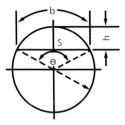

$$A = \frac{\pi \times r^2 \times \Theta}{360} - \frac{S(r - h)}{2}$$

$$S = 2r \sin \frac{\Theta}{2} \qquad h = \frac{S}{2} \tan \frac{\Theta}{4}$$

Approximate area of segment $\quad h = \frac{2}{3} \times S \times h$

---

## Ellipse

$$A = \frac{\pi}{4} \times \text{Major axis} \times \text{Minor axis}$$

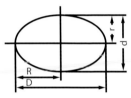

$$A = \frac{\pi}{4} \times D \times d \qquad A = .0785 \times D \times d \qquad A = \pi \times R \times \gamma$$

Circumference C depends on ratio d: D

| Radio d : D | C = D times | d : D | C = D times |
|---|---|---|---|
| 0.9 | 2.9866 | 0.5 | 2.4221 |
| 0.8 | 2.8361 | 0.4 | 2.3013 |
| 0.7 | 2.6912 | 0.3 | 2.1930 |
| 0.6 | 2.5527 | 0.2 | 2.1010 |

Example
D = 150mm
d= 90mm
d:D =90: 150 = 0.6
C = 150 × 2.5527
C = 382.9 mm

## Prism

Volume = Base × Height

$$V = A \times h$$

$V = A_\square \times h$
$V = A_\square \times h$
$V = l \times b \times h$

Surface area for square base

$$A_O = 2A_\square + 4A_\square$$

= 2 × Area of base +
  4 × Area of each side
= 2 × l + b + 4 × l × h

For Rectangle base

$$A_O = 2A_\square + 2A_{\square 1} + 2A_{\square 2}$$

## Prism

$$V = A \times h$$

$V = A_\triangle \times h$

$$V = \frac{l \times h}{2} \times h$$

$$A_O = 2A_\triangle + 3A_{\square 1}$$

A = Equilateral triangle base

$$A_O = 2A_\triangle + A_{\square 1} + A_{\square 2} + A_{\square 3}$$

A = Scalene triangle base

$$A_O = 2A_\triangle + n \times A_\square$$

Regular polygon of n sides

## Cylinder

$$V = A \times h$$

$V = A_O \times h$

$$V = \frac{\pi}{4} \times D^2 \times h$$

Lateral area of the cylinder

$$A_L = \pi \times D \times h$$

## Pyramid

Volume = $\dfrac{\text{Base} \times \text{Height}}{3}$

$$V = \frac{A \times h}{3}$$

$$V = \frac{A_\square \times h}{3}$$

$$V = \frac{l \times b \times h}{3}$$

Surface

$$A_O = 2A_\square + 4A_{\triangle 1}$$

Square base

$$A_O = A_\square + 2A_{\triangle 1} + 2A_{\triangle 2}$$

Rectangle base

## Pyramid

$$V = \frac{A \times h}{3}$$

$$V = \frac{A_\triangle \times h}{3}$$

$$V = \frac{l \times h' \times h}{3}$$

$$A_O = A_\triangle + 3A_{\triangle 1}$$

Equilateral triangle base

$$A_O = A_\triangle + A_{\triangle 1} + A_{\triangle 2} + A_{\triangle 3}$$

Scalene triangle base

$$A_O = A_\triangle + n \times A_{\triangle 1}$$

Regular polygon of n sides

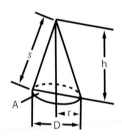

## Cone

$$V = \frac{A \times h}{3}$$

$$V = A_O \times h$$

$$V = \frac{\pi}{4} \times D^2 \times \frac{h}{3}$$

$$A_O = \frac{\pi}{4} D^2 (D + 2S)$$

Lateral surface

$$A_L = \frac{\pi \times D \times S}{2}$$

$$A_L = \pi \times r \times \sqrt{r^2 + h^2}$$

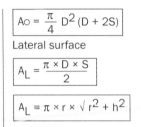

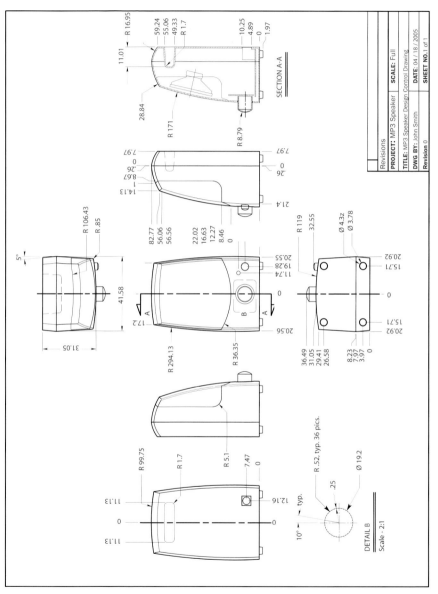

Revisions
**PROJECT:** MP3 Speaker   **SCALE:** Full
**TITLE:** MP3 Speaker Design Control Drawing
**DWG BY:** John Smith   **DATE:** 04 / 18 / 2005
**Revision** 0   **SHEET NO.** 1 of 1

SECTION A-A

DETAIL B
Scale - 2:1

Example of a complete design control drawing, including section and detail views and dimensions.

## 3D MODELING

Up to this point, we have been discussing 2D (two-dimensional) drafting as a means to define and document a design, which at some point needs to be translated into a three-dimensional product. This conversion for many years occurred at the tooling stage. Either a pattern was made for some tooling methods, or the toolmaker cut the 3D interpretation into the mold for production parts. As 3D computer modeling was developed, manufacturing centers and engineering began working in 3D, which helped solve interpretation problems and drawing errors before a mold was made. 3D modeling moved into the industrial design arena after this and is quickly becoming a standard process for design documentation.

### Three-Dimensional Digital Files

3D modeling software for industrial design is not only a tool to help the designer develop the final product but also serves as the digital document to communicate with engineering and others. The key difference between some of the software programs is the type of file that can be created: wire-frame, surfaces, or solids. In addition to modeling, many programs also offer advanced computer rendering capabilities to help present a virtually photo-realistic image. This option is helpful for presenting the final design for business and marketing purposes. It can also reduce the need for extra physical models, as variations in finish can be presented in a digital format. For the primary purpose of design documentation, we will focus on the 3D file, how it is created, and what information is transferred from design to engineering and manufacturing in this document.

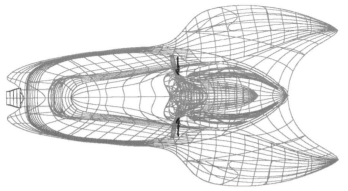

Courtesy of Paradigm Design Associates

Wire frame models represent the underlying design structure of a 3D model. Few programs work only in wire frame. This is, however, a common working and viewing option in most 3D software. This viewing mode allows faster work due to the reduced computations required for complex surface shading.

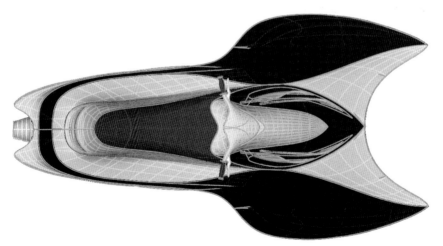

Courtesy of Paradigm Design Associates

Surface models, just as the name implies, represent the outer skin (surface) of the object. These can used to describe the intended form for design and modeling purposes.

Courtesy of Paradigm Design Associates

Solid models, also known as volume models, represent the entire solid object, not just the surface. This includes interior details and wall thicknesses. Assembly drawings containing several components can be put together to verify clearances, and animations can be developed in many programs to check clearances of movable parts. A solid model can be used for visual verification, engineering, rapid prototyping, and, ultimately, manufacturing.

**Communication across multiple platforms**  With the growth of this market creating many competitors running on different machines, one of the challenges has been to maintain communication between systems as a project develops. This very issue has forced many corporations to establish global standards on both hardware and software to work with in new product development. Even different versions of the same software can create a problem in transferring files to one another.

**Data exchange and software variations**  Working on a common platform may simplify some things in a corporate environment, but is not the only way to work together as a group. Data exchange through neutral file formats has been developed so that files can be converted to a basic computer code that can be read by many software programs working on different systems. The translation format must be designed into the software itself. A few common data exchange formats include IGES, DXF, STEP, VDA/FS, VDA/IS, and DES. The process usually involves "exporting" a file out from the original program as an exchange-formatted file, e.g. "mp3speaker.dxf." This file can then be imported into a new software program. The conversion takes place during the export and import functions.

**Document management** In addition to basic communication/translation issues addressed with data exchange, file management is another issue that needs to be addressed as it relates to design documentation and how to insure design integrity through production. With hand drawings, the use of title boxes with names, dates, and revisions help ensure any member of the team is following the most recent version of the design, wherever the member does the work. With digital files, often being transferred through email, it can become more difficult to track the current version of a design with multiple versions existing around the globe.

**FTP: File Transfer Protocol** Using FTP servers over the internet can help manage and control communication of a design by keeping a central file, which can be accessed and downloaded by authorized personnel and updated at a single source. FTP file management can work in two primary ways:

**1.** To use the system to download the current file. Any work done on the file will always need to be sent back to a source manager so that he/she updates the site. This must only be handled at one location.

**2.** To establish an interface over the web so that authorized users can manipulate the file at the FTP source and leave the updated file on the FTP server.

This maintains a single up-to-date file rather than sending updated files around the globe with every revision. This also creates a process for anyone associated with a project to verify and use the current document.

# Chapter 10: Graphics and Type

by Amy Nehez-Cuffaro

## OVERVIEW

The presentation of ideas and information is often a critical component in the product development process. A well-orchestrated visual presentation can create context for an idea, represent intangibles, and explain the details of a concept or concepts. Because page layout and image creation are so easily handled with computers, it is essential for the industrial designer to have a rudimentary knowledge of graphic design terminology, page layout, and file preparation. *The Graphic Design Reference + Specification Book* (Rockport, 2013) is an excellent reference for graphic design terminology.

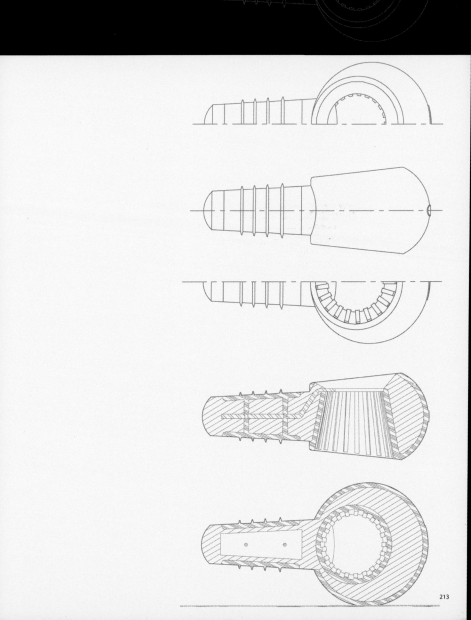

## TYPOGRAPHY TERMS

Because type is measured and described in a language that is unique to the printed word, it is important to recognize and understand basic typographic and typesetting terms.

**Gutter** The white space between columns of type or between pages on a two-page spread

**Kerning/letterspacing/tracking** Adjusting the amount of space between letters or characters so that letter spacing appears to be in balance

**Leading** The amount of vertical space between lines of type

"I have a dream that one day this nation will rise up and live out the true meaning of its creed. 'We hold these truths to be self-evident: that all men are created equal.' I have a dream that one day on the red hills of Georgia the sons of former slaves and the sons of former slaveowners will be able to sit down together at a table of brotherhood. I have a dream that one day even the state of Mississippi, a desert state, sweltering with the heat of injustice and oppression, will be transformed into an oasis of freedom and justice. I have a dream that my four children will one day live in a nation where they will not be judged by the color of their skin but by the content of their character. I have a dream today."

6 pt   **E**

8 pt   **E**

9 pt   **E**

10 pt   **E**

12 pt   **E**

14 pt   **E**

18 pt   **E**

20 pt   **E**

**Point size**   A typographic measurement system used for measuring the height of type, thickness of rules, and leading.

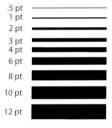

.5 pt
1 pt
2 pt
3 pt
4 pt
6 pt
8 pt
10 pt
12 pt

24 pt   **E**

30 pt   **E**

36 pt   **E**

42 pt   **E**

48 pt   **E**

54 pt   **E**

60 pt   **E**

The E-scale can be used to estimate the point size of a typeface.

72 pt   **E**

Uppercase Letter · Serif · Lowercase Letter · Ascender · Cap Height · Baseline · Serif · Descender

Serif Type · Sans-Serif Type

**Serif typefaces** Faces that originated with the Romans who identified their stone shrines and public buildings with chisel-cut letterforms. To hide the ragged ends of those letterforms, the Romans would cut a short extra stroke on the ends of their letters. This extra cut was a called a serif.

**Sans-serif typefaces** Born out of the Industrial Revolution to reflect a more modern aesthetic, they are characterized by no serifs and a smooth, streamlined look.

**Typeface** The design of a single set of letterforms, numerals, and punctuation marks unified by consistent visual properties. Typeface designs are identified by name, such as Times or Helvetica.

**Type font** A set of characters in a specific typeface, at a specific point size, and in a specific style. 11-point Helvetica Bold is a font—the typeface, at 11-point size, in the bold style. Hence, 12-point Helvetica Italic and 9-point Bold are separate fonts.

**Type style** Modifications in a typeface that create design variety while maintaining the visual character of the typeface. These include variations in weight (light, medium, book, bold, heavy, and black), width (condensed or extended), or angle (italic or oblique vs. roman or upright).

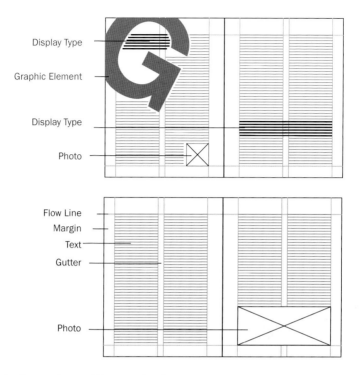

This example of a two-column grid is appropriate for a standard size double-sided magazine spread, vertical format.

**Widow**  A word or part of a word that is the last line of a paragraph or that ends up at the top of a page by itself. Widows are considered unattractive and inhibit readability in a layout.

**Grid**  The grid is the invisible framework on which a page or composition of elements is designed. This essential framework divides the page into a modular structure of horizontals and verticals. The grid gives the composition consistency and is the glue that holds all the visual elements together.

The grid is intended to be a guide, and the elements on the page can interrupt the grid with careful consideration. The scale and proportion of columns, margins, and visual elements and their relationship to one another must be carefully adjusted. The underlying principle is one of ordered unity.

**Color Systems** Color, when viewed on a computer screen, can look very different from a printed version of the same color. In fact, no amount of adjusting or calibrating a computer monitor can yield a truly accurate representation of how a color will look when it is printed. This difference occurs because digital color is expressed as projected light—a combination of red, green, and blue (RGB). Printed, or reflected, color is produced as a combination of cyan, magenta, yellow, and black (CMYK).

Besides being able to specify combinations and percentages of cyan, magenta, yellow, and black (CMYK process), there are several color systems from which to choose. Below is a sample of the most commonly used specifying systems. To get accurate on-press color, find the color swatch that matches the color you would like to create on press, and then specify that color to the printer.

**Pantone Matching System** Used for printing inks. A Pantone color is selected from a Pantone Color Guide and is specified by a PMS number. Each color has a specified CMYK equivalent. If a job calls for "four-color process," a Pantone color can be chosen and then converted to a process color by specifying the CMYK equivalent of that Pantone color. However, the final CMYK equivalent may be slightly different.

**Toyo** Consists of more than 1,000 colors based on the most common printing inks used in Japan.

**Munsell** Created by Albert Munsell, the Munsell color system is accepted by many standards and professional organizations concerned with color. The system describes color using three variables: hue, saturation, and value. In the three-dimensional space defined by this system, the central axis represents value, hues are organized around the axis, and saturation increases away from the axis.

---

## Image File Formats

Digital files can be saved in a number of file formats that will allow them to be moved between different types of imaging and page layout applications. These file formats also allow imagery to be cross platformed.

**EPS or .eps** (Encapsulated PostScript) Used for placing images or graphics in documents created in word processing, page layout, or drawing programs. EPS files can be cross platformed, cropped, or edited.

**GIF or .gif** (Graphics Interchange Format) An 8-bit, low-memory option for posting images online, GIF images are limited to 256 colors, making them unsuitable for most print applications. However, their limited color quality makes them ideally suited for the limited color display range of computer monitors. GIFs are well suited for images containing large, flat areas of one color and are often used for graphics such as logos and line art.

**JPEG or .jpg (Joint Photographic Experts Group) A file** format designated by the Joint Photographic Experts Group for image compression. Because it is a "lossy" compression format, image quality is sacrificed to conserve disk space. JPEGs are frequently used for placing imagery in websites and online applications where high-resolution files are not necessary. JPEGs work best for photographs, illustrations, and other complex imagery.

**PDF or .pdf** (Portable Document Format) Used for allowing documents to be viewed and printed independent of the application used to create them. Often used for transferring printed pages over the web, either for downloading existing publications or for sending documents to service bureaus or commercial printers for output.

**TIFF or .tif** (Tagged Image File Format) Used for placing images or graphics in documents created in word processing, page layout, or drawing programs. Supports rasterized data and converts vectored images to bits. TIFF files can be cropped or edited. They are similar to EPS, but smaller file size saves memory over EPS format.

**Images/Scanning**

The quality of the image that you input will determine the quality of the reproduction that you output. When scanning any image for output or reproduction, the general rule is: scan your image as closely to size as possible, varying only 10–15% of its output size. Also, scan the image at twice the dots per inch (DPI) as it is to be reproduced in line screen for output. For example: If a magazine prints at a 150-line screen, the image should be scanned at 300 dots per inch.

# Chapter 11: Measurements and Conversions

### by Daniel F. Cuffaro

## OVERVIEW

The global nature of product development necessitates an understanding of the use and conversions of common systems of measurement. This chapter contains an overview of various measurement systems, frequently used conversions, and comprehensive conversion charts.

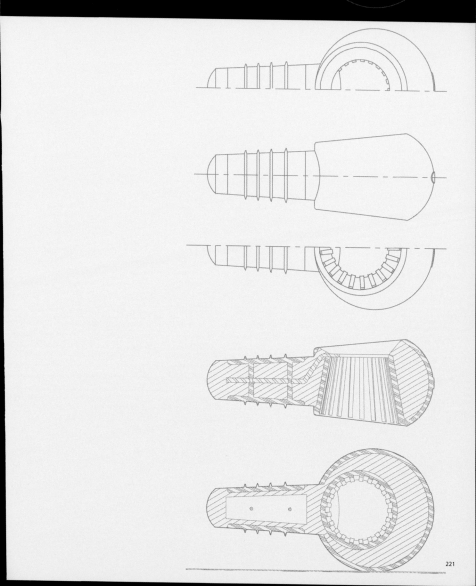

## WEIGHT

| metric | abbreviation | number of gram | U.S. equivalent (approx) |
|--------|--------------|----------------|--------------------------|
| Metric ton | t | 1,000,000 | 1.102 tons |
| Quintal | | 100,000 | 220.462 pounds |
| Kilogram | kg | 1,000 | 2.205 pounds |
| Hectogram | hg | 100 | 3.527 ounces |
| Decagram | dag | 10 | 0.353 ounce |
| Gram | g | 1 | 0.035 ounce |
| Decigram | dg | .10 | – |
| Centigram | cg | .01 | – |
| Milligram | mg | .001 | Length |

| U.S. | abbreviation | common conversion | metric equivalent (approx) |
|------|--------------|-------------------|----------------------------|
| Ton (short) | | 2,000 lbs | .90718 t |
| Pound | lb or # | 16 oz | .454 kg |
| Ounce | oz | 0625 lbs | 20.350 g |

## LENGTH

| metric | abbreviation | number of meters | U.S. equivalent (approx) |
|--------|--------------|------------------|--------------------------|
| Kilometer | km | 1,000 | .62 mile |
| Hectometer | hm | 100 | 109.36 yards |
| Decameter | dam | 10 | 32.81 feet |
| Meter | m | 1 | 39.37 inches |
| Decimeter | dm | .1 | 3.94 inches |
| Centimeter | cm | .01 | .39 inch |
| Millimeter | mm | .001 | .039 inch |

| U.S. | abbreviation | common conversion | metric equivalent (approx) |
|---|---|---|---|
| Mile | mi | 5280 ft | 1.609 km |
| Yard | yd | 3 ft or 36 in | .914 m |
| Foot | ft or ' | 12 in or .333 yd | 30.48 cm |
| Inch | in or " | .083 ft or .028 yd | 2.54 cm |

## CAPACITY

| metric | abbreviation | number of liters | U.S. equivalent (approx) |
|---|---|---|---|
| Kiloliter | kl | 1,000 | 1.308 cubic yards |
| Hectoliter | hl | 100 | 3.532 cubic feet |
| Decaliter | dal | 10 | 0.353 cubic foot |
| Liter | l | 1 | 61.024 cubic inches |
| Cubic decimeter | dm3 | 1.0 | 61.021 cubic inches |
| Deciliter | dl | .10 | 6.102 cubic inches |
| Centiliter | cl | .01 | 0.610 cubic inch |
| Milliliter | ml | .001 | 0.061 cubic inch |

| U.S. (liquid) | abbreviation | common conversion | metric equivalent (approx) |
|---|---|---|---|
| Gallon | gal | 4 qts (231 cu in) | 3.785 liters |
| Quart | qt | 2 pts (57.75 cu in) | .946 liter |
| Pint | pt | (28.875 cu in) | .473 liter |
| Fluid ounce | fl oz | (1.805 cu in) | 29.573 mliters |

| U.S. (dry) | abbreviation | common conversion | metric equivalent (approx) |
|---|---|---|---|
| Bushel | bu | 4 pks (2150.42 cu in) | 35.239 liters |
| Peck | pk | 8 qts (537.605 cu in) | 8.810 liters |
| Quart | qt | 2 pts (67.201 cu in) | 1.101 liters |
| Pint | pt | .5 qt (33.6 cu in) | .0551 liter |

| imperial | abbreviation | common conversion | metric equivalent (approx) |
|----------|--------------|-------------------|----------------------------|
| Bushel | bu | 4 pks (2219.36 cu in) | .035 cu m |
| Peck | pk | 2 gals (554.84 cu in) | .009 cu m |
| Gallon | gal | 4 qt (277.420 cu in) | 4.546 liters |
| Quart | qt | 2 pts (69.355 cu in) | 1.136 liters |
| Pint | pt | (34.678 cu in) | 568.26 cu cm |

## AREA

| metric | abbreviation | number of sq meters | U.S. equivalent (approx) |
|--------|--------------|---------------------|--------------------------|
| Square kilometer | sq km or km2 | 11,000,000 | .3861 square mile |
| Hectare | ha | 10,000 | 2.47 acres |
| Are | a | 100 | 119.6 square yards |
| Square centimeter | sq cm or cm2 | .0001 | .155 square inch |

| U.S. | abbreviation | common conversion | metric equivalent (approx) |
|------|--------------|-------------------|----------------------------|
| Square mile | sq mi or mi2 | 640 acres | 2.590 sq kilometers |
| Acre | | 4840 sq yds | 4046.856 sq meters |
| Square yard | sq yd or yd2 | 27 sq ft | .836 sq meter |
| Square foot | sq ft or ft2 | 144 sq in | .093 sq meter |
| Square inch | sq in or in2 | 0069 sq ft | 6.452 sq cm |

## VOLUME

| metric | abbreviation | number of cu meters | U.S. equivalent (approx) |
|---|---|---|---|
| Cubic centimeter | cu cm, cm3 or cc | .000001 | .061 cu in |
| Cubic decimeter | dm3 | .001 | 61.024 cu in |
| Cubic meter | m3 | 1 | 1.308 cu yds |

| U.S. | abbreviation | common conversion | metric equivalent (approx) |
|---|---|---|---|
| Cubic yard | cu yd or yd3 | 27 cu ft | .765 cu meter |
| Cubic foot | cu ft or ft3 | 1728 cu in | .028 cu meter |
| Cubic inch | cu in or in3 | .00058 cu ft | 16.387 cu cm |

## METRIC SYSTEM (OVERVIEW)

| prefix | abbreviation | multiplier |
|---|---|---|
| Tera | T | 1,000,000,000,000 |
| Giga | G | 1,000,000,000 |
| Mega | M | 1,000,000 |
| Kilo | k | 1,000 |
| Hecto | h | 100 |
| Deca | da | 10 |
| — | — | 1 |
| Deci | d | .1 |
| Centi | c | .01 |
| Milli | m | .001 |
| Micro | μ | .000001 |
| Nano | n | .000000001 |
| Pico | p | .000000000001 |

## STANDARD INCH FRACTIONS TO DECIMAL CONVERSION

| fractions | decimals | fractions | decimals |
|-----------|----------|-----------|----------|
| 1/64 | .015625 | 33/64 | .515625 |
| 1/32 | .03125 | 17/32 | .53125 |
| 3/64 | .046875 | 35/64 | .546875 |
| 1/16 | .0625 | 9/16 | .5625 |
| 5/64 | .078125 | 37/64 | .578125 |
| 3/32 | .09375 | 19/32 | .59375 |
| 7/64 | .109375 | 39/64 | .609375 |
| 1/8 | .125 | 5/8 | .625 |
| 9/64 | .140625 | 41/64 | .640625 |
| 5/32 | .15625 | 21/32 | .65625 |
| 11/64 | .171875 | 43/64 | .671875 |
| 3/16 | .1875 | 11/16 | .6875 |
| 11/64 | .203125 | 45/64 | .703125 |
| 7/32 | .21875 | 23/32 | .71875 |
| 15/64 | .234375 | 47/64 | .734375 |
| 1/4 | .25 | 3/4 | .75 |
| 17/64 | .265625 | 49/64 | .765625 |
| 9/32 | .28125 | 25/32 | .78125 |
| 19/64 | .296875 | 51/64 | .796875 |
| 5/16 | .3125 | 13/16 | .8125 |
| 21/64 | .328125 | 53/64 | .828125 |
| 11/32 | .34375 | 27/32 | .84375 |
| 23/64 | .359375 | 55/64 | .859375 |
| 3/8 | .375 | 7/8 | .875 |
| 25/64 | .390625 | 57/64 | .890625 |
| 13/32 | .40625 | 29/32 | .90625 |
| 27/64 | .421875 | 59/64 | .921875 |

| fractions | decimals | fractions | decimals |
|-----------|----------|-----------|----------|
| 7/16 | .4375 | 15/16 | .9375 |
| 29/64 | .453125 | 61/64 | .953125 |
| 15/32 | .46875 | 31/32 | .96875 |
| 31/64 | .484375 | 63/64 | .984375 |
| 1/2 | .5 | 1 | 1. |

## TYPE CONVERSIONS

| inches | mm | points | picas |
|--------|-----|--------|-------|
| 1/16 | 1.588 | 4.50 | 0.375 |
| 1/8 | 3.175 | 9.0 | .75 |
| 3/16 | 4.763 | 13.5 | 1.125 |
| 1/4 | 6.35 | 18.0 | 1.5 |
| 5/16 | 7.938 | 22.5 | 1.875 |
| 3/8 | 9.525 | 27.0 | 2.25 |
| 7/16 | 11.113 | 31.5 | 2.625 |
| 1/2 | 12.7 | 36.0 | 3.0 |
| 9/16 | 14.288 | 40.5 | 3.375 |
| 5/8 | 15.875 | 45.0 | 3.75 |
| 11/16 | 17.463 | 49.5 | 4.125 |
| 3/4 | 19.050 | 54.0 | 4.5 |
| 13/16 | 20.638 | 58.5 | 4.875 |
| 7/8 | 22.225 | 63.0 | 5.25 |
| 14/16 | 23.813 | 67.5 | 5.625 |
| 1 | 25.4 | 72.0 | 6.0 |

| multiply | by | to obtain |
|---|---|---|
| Acre-feet | 43,560 | Cubic feet |
| Acre-feet | 1233.48 | Cubic meters |
| Acre-feet | 325,851 | Gallons |
| Acres | 0.4047 | Hectares |
| Acres | 43,560 | Square feet |
| Acres | 4,046.86 | Square meters |
| Acres | 0.001562 | Square miles |
| Acres | 4,840 | Square yards |
| Ares | 0.02471 | Acres |
| Ares | 100 | Square meters |
| | | |
| Bags (cement) | 94 | Pounds |
| Barrels (cement) | 376 | Pounds |
| Barrels (dry) | 7,056 | Cubic inches |
| Barrels (dry) | 105 | Quarts (dry) |
| Barrels (liquid) | 31.5 | Gallons |
| Barrels (oil) | 42 | Gallons |
| Board feet | 144 | Cubic inches |
| Bolts (US Cloth) | 40 | Yards |
| Buckets (British, Dry) | 18.18 | Liters |
| Bushels | 1.2445 | Cubic feet |
| Bushels | 2150.4 | Cubic inches |
| Bushels | 0.03524 | Cubic meters |
| Bushels | 35.239 | Liters |
| Bushels | 4 | Pecks |
| Bushels | 64 | Pints (dry) |
| Bushels | 32 | Quarts (dry) |

| multiply | by | to obtain |
|---|---|---|
| Centares (Centiares) | 1 | Square meters |
| Centigrams | 0.01 | Grams |
| Centiliters | 0.01 | Liters |
| Centimeter-grams | 0.00001 | Meter-kilograms |
| Centimeter-grams | .00007233 | Pound-feet |
| Centimeters | 0.3281 | feet |
| Centimeters | 0.3937 | Inches |
| Centimeters | 0.01 | Meters |
| Centimeters | 10 | Millimeters |
| Centimeters | 393.7 | Mils |
| Centimeters/second | 1.9685 | Feet/minute |
| Centimeters/second | 0.03281 | Feet/second |
| Centimeters/second | 0.036 | Kilometers/hour |
| Centimeters/second | 0.6 | Meters/minute |
| Centimeters/second | 0.02237 | Miles/hour |
| Centimeters/second | 0.0003728 | Miles/minute |
| Centimeters/sec./sec. | 0.03281 | Feet/sec./sec. |
| Circular mils | 0.000005067 | Square centimeters |
| Circular mils | 0.7854 | Square mils |
| Circular mils | 0.0000007854 | Square inches |
| Circumference | 6.2832 | Radians |
| Cubic centimeters | 0.0000353 | Cubic feet |
| Cubic centimeters | 0.06102 | Cubic inches |
| Cubic centimeters | 0.000001 | Cubic meters |
| Cubic centimeters | 0.000001308 | Cubic yards |
| Cubic centimeters | 0.00106 | Cups |
| Cubic centimeters | 0.0002642 | Gallons |
| Cubic centimeters | 0.001 | Liters |

| multiply | by | to obtain |
| --- | --- | --- |
| Cubic centimeters | 0.002113 | Pints (liquid) |
| Cubic centimeters | 0.001057 | Quarts (liquid) |
| Cubic feet | 0.8036 | Bushels (dry) |
| Cubic feet | 28,316.8 | Cubic centimeters |
| Cubic feet | 1728 | Cubic inches |
| Cubic feet | 0.02832 | Cubic meters |
| Cubic feet | 0.03704 | Cubic yards |
| Cubic feet | 7.48052 | Gallons |
| Cubic feet | 28.317 | Liters |
| Cubic feet | 59.844 | Pints (liquid) |
| Cubic feet | 62.43 | Pounds of water at 39.2°F |
| Cubic feet | 29.922 | Quarts (liquid) |
| Cubic feet/minute | 471.95 | Cubic centimeters/second |
| Cubic feet/minute | 0.1247 | Gallons/second |
| Cubic feet/minute | 0.472 | Liters/second |
| Cubic feet/minute | 62.43 | Pounds of water/minute |
| Cubic feet/second | 0.646317 | Millions of gallons/day |
| Cubic feet/second | 448.831 | Gallons/minute |
| Cubic inches | 16.371 | Cubic centimeters |
| Cubic inches | 0.0005787 | Cubic feet |
| Cubic inches | 0.00001639 | Cubic meters |
| Cubic inches | 0.00002143 | Cubic yards |
| Cubic inches | 0.004329 | Gallons |
| Cubic inches | 0.01639 | Liters |
| Cubic inches | 106,100 | Mil-feet |
| Cubic inches | 0.03463 | Pints (liquid) |
| Cubic inches | 0.01732 | Quarts (liquid) |
| Cubic meters | 28.38 | Bushels (dry) |

| multiply | by | to obtain |
|---|---|---|
| Cubic meters | 1,000,000 | Cubic centimeters |
| Cubic meters | 35.3147 | Cubic feet |
| Cubic meters | 61,023.7 | Cubic inches |
| Cubic meters | 1.308 | Cubic yards |
| Cubic meters | 264.172 | Gallons |
| Cubic meters | 1,000 | Liters |
| Cubic meters | 2,113 .38 | Pints (liquid) |
| Cubic meters | 1,056.69 | Quarts (liquid) |
| Cubic yards | 764,554.9 | Cubic centimeters |
| Cubic yards | 27 | Cubic feet |
| Cubic yards | 46,656 | Cubic inches |
| Cubic yards | 0.7646 | Cubic meters |
| Cubic yards | 201.974 | Gallons |
| Cubic yards | 764.555 | Liters |
| Cubic yards | 1,615.79 | Pints (liquid) |
| Cubic yards | 807.896 | Quarts (liquid) |
| Cubic yards/minute | 0.45 | Cubic feet/second |
| Cubic yards/minute | 3.367 | Gallons/second |
| Cubic yards/minute | 12.742 | Liters/second |
| Cups | 23.6588 | Cubic centimeters |
| Cups | 0.0625 | Gallons |
| Cups | 0.2366 | Liters |
| Cup | 8 | Ounces |
| Cups | 0.5 | Pints (liquid) |
| Cups | 0.25 | Quarts (liquid) |
| Cups | 16 | Tablespoons |
| Cups | 48 | Teaspoons |

| multiply | by | to obtain |
| --- | --- | --- |
| Days | 86,400 | Seconds |
| Decigrams | 0.1 | Grams |
| Deciliters | 0.1 | Liters |
| Decimeters | 0.1 | Meters |
| Degrees (angle) | 60 | Minutes |
| Degrees (angle) | 0.01111 | Quadrants |
| Degrees (angle) | 0.01745 | Radians |
| Degrees (angle) | 3,600 | Seconds |
| Degrees (Celsius) | (°Cx9/5)+32 | Degrees (Fahrenheit) |
| Degrees (Celsius) | °C+273.15 | Degrees (Kelvin) |
| Degrees (Fahrenheit) | (°F-32)x5/9 | Degrees (Celsius) |
| Degrees (Fahrenheit) | °F+459.67 | Degrees (Rankine) |
| Degrees (Kelvin) | °K-273.15 | Degrees (Celsius) |
| Degrees (Rankine) | °R-459.67 | Degrees (Fahrenheit) |
| Degrees/second (angle) | 0.01745 | Radians/second |
| Degrees/second (angle) | 0.1667 | Revolutions/minute |
| Degrees/second (angle) | 0.002778 | Revolutions/second |
| Decagrams | 10 | Grams |
| Decaliters | 10 | Liters |
| Decameters | 10 | Meters |
| Em (Pica) | 0.167 | Inches |
| Em (Pica) | 0.4233 | Centimeters |
| Feet | 30.48 | Centimeters |
| Feet | 12 | Inches |
| Feet | 0.0003048 | Kilometers |
| Feet | 0.3048 | Meters |

| multiply | by | to obtain |
|---|---|---|
| Feet | 0.000189 | Miles |
| Feet | 0.0001645 | Miles (nautical) |
| Feet | 304.8 | Millimeters |
| Feet | 0.3333 | Yards |
| Feet/minute | 0.508 | Centimeters/second |
| Feet/minute | 0.01667 | Feet/second |
| Feet/minute | 0.01829 | Kilometers/hour |
| Feet/minute | 0.3048 | Meters/minute |
| Feet/minute | 0.01136 | Miles/hour |
| Feet/second | 30.48 | Centimeters/second |
| Feet/second | 1.0973 | Kilometers/hour |
| Feet/second | 0.5921 | Knots |
| Feet/second | 18.288 | Meters/minute |
| Feet/second | 0.6818 | Miles/hour |
| Feet/second | 0.01136 | Miles/minute |
| Feet/sec./sec. | 30.48 | Centimeters/sec./sec. |
| Feet/sec./sec. | 0.3048 | Meters/sec./sec. |
| Feet/sec./sec. | 1.0973 | Kilometers/hr./sec. |
| Feet/sec./sec. | 0.6818 | Miles/hour/sec. |
| g's (gravity) | 32.174 | feet/sec/sec |
| g's (gravity) | 9.807 | meters/sec/sec |
| Gallons | 3,785.41 | Cubic centimeters |
| Gallons | 0.1337 | Cubic feet |
| Gallons | 231 | Cubic inches |
| Gallons | 0.003785 | Cubic meters |
| Gallons | 0.004951 | Cubic yards |
| Gallons | 16 | Cup |

| multiply | by | to obtain |
|---|---|---|
| Gallons | 0.83267 | Gallons (Imperial) |
| Gallons | 3.7854 | Liters |
| Gallons | 128 | Ounces |
| Gallons | 8 | Pints (liquid) |
| Gallons | 8.3371 | Pounds of water at 60° F |
| Gallons | 4 | Quarts (liquid) |
| Gallons | 256 | Tablespoon |
| Gallons | 768 | Teaspoon |
| Gallons (Imperial) | 1.20095 | Gallons |
| Gallons/minute | 8.0208 | Cubic feet/hour |
| Gallons/minute | 0.002228 | Cubic feet/second |
| Gallons/minute | 0.06309 | Liters/second |
| Gallons/minute | 6.0086 | Tons of water/24 hrs. |
| Grade percentage | 0.9001 | Degrees |
| Grams | 0.001 | Kilograms |
| Grams | 1000 | Milligrams |
| Grams | 0.03527 | Ounces |
| Grams | 0.03215 | Ounces (troy) |
| Grams | 0.002205 | Pounds |
| Grams/centimeter | 0.0056 | Pounds/inch |
| Grams/cubic centimeter | 62.428 | Pounds/cubic foot |
| Grams/cubic centimeter | 0.03613 | Pounds/cubic inch |
| Grams/liter | 8.3454 | Pounds/1000 gallons |
| Grams/liter | 0.062427 | Pounds/cubic foot |
| Grams/liter | 1,000 | Parts/million |
| Grams-force/square centimeter | 2.0481 | Pounds-force/square foot |

| multiply | by | to obtain |
|---|---|---|
| Hands | 4 | Inches |
| Hectares | 2.471 | Acres |
| Hectares | 10,000 | Square meters |
| Hectares | 107,639.1 | Square feet |
| Hectograms | 100 | Grams |
| Hectoliters | 100 | Liters |
| Hectometers | 100 | Meters |
| Horsepower | 33,000 | Foot-pounds/minute |
| Horsepower | 550 | Foot-pounds/second |
| Horsepower | 1.014 | Horsepower (metric) |
| Horsepower | 0.7457 | Kilowatts |
| Horsepower | 745.7 | Watts |
| Horsepower (metric) | 0.9863 | Horsepower |
| Inches | 2.54 | Centimeters |
| Inches | 0.0254 | Meters |
| Inches | 25.4 | Millimeters |
| Inches | 1,000 | Mils |
| Inches | 0.02778 | Yards |
| Kilogram-meters/ sec./sec. | 1 | Newtons |
| Kilograms | 1,000 | Grams |
| Kilograms | 2.2046 | Pounds |
| Kilograms | 0.001102 | Tons |
| Kilograms/cubic meter | 0.001 | Grams/cubic centimeter |
| Kilograms/cubic meter | 0.06243 | Pounds/cubic foot |
| Kilograms/cubic meter | 0.00003613 | Pounds/cubic inch |

| multiply | by | to obtain |
|----------|-----|-----------|
| Kilograms/meter | 0.672 | Pounds/foot |
| Kilograms/sq. centimeter | 2,048 | Pounds/square foot |
| centimeter | 14.223 | Pounds/square inch |
| Kilograms/sq. meter | 0.2048 | Pounds/square foot |
| Kilograms/sq. meter | 0.001422 | Pounds/square inch |
| Kilograms/sq. millimeter | 1,000,000 | Kilograms/square meter |
| Kiloliters | 1,000 | Liters |
| Kilometers | 100,000 | Centimeters |
| Kilometers | 3,280.84 | Feet |
| Kilometers | 1,000 | Meters |
| Kilometers | 0.6214 | Miles |
| Kilometers | 1,093.61 | Yards |
| Kilometers/hour | 27.78 | Centimeters/second |
| Kilometers/hour | 54.68 | Feet/minute |
| Kilometers/hour | 0.9113 | Feet/second |
| Kilometers/hour | 0.53996 | Knots |
| Kilometers/hour | 16.668 | Meters/minute |
| Kilometers/hour | 0.6214 | Miles/hour |
| Kilometers/hour/sec. | 27.78 | Centimeters/sec./sec. |
| Kilometers/hour/sec. | 0.9113 | Feet/sec./sec. |
| Kilometers/hour/sec. | 0.2778 | Meters/sec./sec. |
| Knots (Int) | 1.8532 | Kilometers/hr. |
| Knots (Int) | 1 | Nautical miles/hour |
| Knots (Int) | 1.1508 | Miles/hour |
| Knots (Int) | 1.689 | Feet/second |

| multiply | by | to obtain |
|---|---|---|
| Leagues (Statute) | 3 | Miles |
| Light years | 9.46073x1012 | Kilometers |
| Light years | 5.8785x1012 | Miles |
| Links (engineer's) | 12 | Inches |
| Links (surveyor's) | 7.92 | Inches |
| Liters | 1,000 | Cubic centimeters |
| Liters | 0.03531 | Cubic feet |
| Liters | 61.0237 | Cubic inches |
| Liters | 0.001 | Cubic meters |
| Liters | 0.001308 | Cubic yards |
| Liters | 4.22675 | Cup |
| Liters | 0.2642 | Gallons |
| Liters | 1 | Kilograms of water at 39.2°F |
| Liters | 33.81402 | Ounce |
| Liters | 2.1134 | Pints (liquid) |
| Liters | 1.0567 | Quarts (liquid) |
| Liters | 67.628451 | Tablespoon |
| Liters | 202.884135 | Teaspoon |
| Liters/minute | 0.0005886 | Cubic feet/second |
| Liters/minute | 0.004403 | Gallons/second |
| Meter-kilograms | 1,000,000 | Centimeter-grams |
| Meter-kilograms | 7.233 | Pound-feet |
| Meters | 100 | Centimeters |
| Meters | 3.2808 | Feet |
| Meters | 39.3701 | Inches |
| Meters | 0.001 | Kilometers |
| Meters | 0.000621 | Miles |

| multiply | by | to obtain |
| --- | --- | --- |
| Meters | 0.000539 | Miles (nautical) |
| Meters | 1,000 | Millimeters |
| Meters | 1.0936 | Yards |
| Meters/minute | 1.667 | Centimeters/second |
| Meters/minute | 0.032397 | Knots |
| Meters/minute | 3.2808 | Feet/minute |
| Meters/minute | 0.05468 | Feet/second |
| Meters/minute | 0.06 | Kilometers/hour |
| Meters/minute | 0.03728 | Miles/hour |
| Meters/second | 196.8 | Feet/minute |
| Meters/second | 3.2808 | Feet/second |
| Meters/second | 3.6 | Kilometers/hour |
| Meters/second | 0.06 | Kilometers/minute |
| Meters/second | 2.2369 | Miles/hour |
| Meters/second | 0.03728 | Miles/minute |
| Meters/sec./sec. | 3.281 | Feet/sec./sec. |
| Meters/sec./sec. | 3.6 | Kilometers/hr./sec. |
| Meters/sec./sec. | 2.2369 | Miles/hour/sec. |
| Microns | 0.000009425 | Cu. inches |
| Miles | 160,344 | Centimeters |
| Miles | 5,280 | Feet |
| Miles | 63,360 | Inches |
| Miles | 1.609 | Kilometers |
| Miles | 1,609.344 | Meters |
| Miles | 0.86898 | Miles (nautical) |
| Miles | 1,760 | Yards |
| Miles (nautical) | 6076.12 | Feet |
| Miles (nautical) | 1.852 | Kilometers |

| multiply | by | to obtain |
|---|---|---|
| Miles (nautical) | 1852 | Meters |
| Miles (nautical) | 1.15078 | Miles |
| Miles (nautical) | 2,025.373 | Yards |
| Miles/hour | 44.704 | Centimeters/second |
| Miles/hour | 88 | Feet/minute |
| Miles/hour | 1.4667 | Feet/second |
| Miles/hour | 1.6093 | Kilometers/hour |
| Miles/hour | 0.868976 | Knots |
| Miles/hour | 26.822 | Meters/minute |
| Miles/hour | 0.01667 | Miles/minute |
| Miles/hour/sec. | 44.704 | Centimeters/sec./sec. |
| Miles/hour/sec. | 1.4667 | Feet/sec./sec. |
| Miles/hour/sec. | 1.6093 | Kilometers/hr./sec. |
| Miles/hour/sec. | 0.447 | Meters/sec./sec. |
| Miles/minute | 2,682.2 | Centimeters/second |
| Miles/minute | 88 | Feet/second |
| Miles/minute | 1.6093 | Kilometers/minute |
| Miles/minute | 60 | Miles/hour |
| Milligrams | 0.001 | Grams |
| Milligrams/liter | 1 | Parts/million |
| Milliliters | 0.001 | Liters |
| Millimeters | 0.1 | Centimeters |
| Millimeters | 0.03937 | Inches |
| Millimeters | 0.003281 | Feet |
| Millimeters | 39.37 | Mils |
| Millimicrons | 0.00000001 | Meters |
| Million gallons/day | 1.54723 | Cubic feet/second |
| Mils | 0.00254 | Centimeters |

| multiply | by | to obtain |
|---|---|---|
| Mils | 0.001 | Inches |
| Minutes (angle) | 0.01667 | Degrees |
| Minutes (angle) | 0.0001852 | Quadrants |
| Minutes (angle) | 0.0002909 | Radians |
| Minutes (angle) | 60 | Seconds |
| Newton-meters | 1 | Joules |
| Newtons | 1 | Kilogram-meters/sq sec |
| Newtons | 0.2248089 | Pounds |
| Ounces | 28.349527 | Grams |
| Ounces | 0.91146 | Ounces (troy) |
| Ounces | 0.0625 | Pounds |
| Ounces | 0.0000279 | Tons (long) |
| Ounces | 0.00002835 | Tons (metric) |
| Ounces (fluid) | 29.5735 | Cubic centimeters |
| Ounces (fluid) | 1.80469 | Cubic inches |
| Ounces (fluid) | 0.0078125 | Gallons |
| Ounces (fluid) | 0.02957 | Liters |
| Ounces (fluid) | 0.0625 | Pints |
| Ounces (fluid) | 0.03125 | Quarts |
| Ounces (fluid) | 2 | Tablespoons |
| Ounces (fluid) | 6 | Teaspoons |
| Ounces (troy) | 31.103481 | Grams |
| Ounces (troy) | 1.09714 | Ounces |
| Ounces (troy) | 0.08333 | Pounds (troy) |
| Ounces/square inch | 0.0625 | Pounds/square inch |

| multiply | by | to obtain |
| --- | --- | --- |
| Parts/million | 8.345 | Pounds/million gallons |
| Pecks | 0.25 | Bushels |
| Pecks | 537.605 | Cubic inches |
| Pecks | 8.8098 | Liters |
| Pecks | 8 | Quarts (dry) |
| Pi | 1 | 3.141592654 |
| Pints (dry) | 33.6003 | Cubic inches |
| Pints (liquid) | 473.1765 | Cubic Centimeters |
| Pints (liquid) | 0.01671 | Cubic feet |
| Pints (liquid) | 28.875 | Cubic inches |
| Pints (liquid) | 0.0006189 | Cubic yards |
| Pints (liquid) | 2 | Cups |
| Pints (liquid) | 0.125 | Gallons |
| Pints (liquid) | 0.473176 | Liters |
| Pints (liquid) | 16 | Ounces |
| Pints (liquid) | 0.85934 | Pints |
| Pints (liquid) | 0.5 | Quarts (liquid) |
| Pounds | 453.5924 | Grams |
| Pounds | 0.4536 | Kilograms |
| Pounds - force | 4.44822 | Newtons |
| Pounds | 16 | Ounces |
| Pounds | 14.5833 | Ounces (troy) |
| Pounds | 1.21528 | Pounds (troy) |
| Pounds | 0.0005 | Tons (short) |
| Pounds of water | 0.01602 | Cubic feet |
| Pounds of water | 27.68 | Cubic inches |
| Pounds of water | 0.1198 | Gallons |
| Pounds of water/min. | 0.000267 | Cubic feet/second |

| multiply | by | to obtain |
| --- | --- | --- |
| Pounds/cubic foot | 0.016018 | Grams/cubic centimeter |
| Pounds/cubic foot | 16.018 | Kilograms/cubic meter |
| Pounds/cubic foot | 0.0005787 | Pounds/cubic inch |
| Pounds/cubic inch | 27.6799 | Grams/cubic centimeter |
| Pounds/cubic inch | 27,679.9 | Kilograms/cubic meter |
| Pounds/cubic inch | 1,728 | Pounds/cubic foot |
| Pounds/foot | 1.488 | Kilograms/meter |
| Pounds/inch | 178.5797 | Grams/centimeter |
| Pounds/square foot | 0.01602 | Feet of water |
| Pounds/square foot | 4.88243 | Kilograms/square meter |
| Pounds/square foot | 0.006945 | Pounds/square inch |
| Pounds/square inch | 703.1 | Kilograms/square meter |
| Quadrants (angle) | 90 | Degrees |
| Quadrants (angle) | 5,400 | Minutes |
| Quadrants (angle) | 1.571 | Radians |
| Quarts (dry) | 67.2006 | Cubic inches |
| Quarts (dry) | 1.1636 | Quarts (liquid) |
| Quarts (liquid) | 946.353 | Cubic centimeters |
| Quarts (liquid) | 0.03342 | Cubic feet |
| Quarts (liquid) | 57.75 | Cubic inches |
| Quarts (liquid) | 0.0009464 | Cubic meters |
| Quarts (liquid) | 4 | Cups |
| Quarts (liquid) | 0.25 | Gallons |
| Quarts (liquid) | 0.94635 | Liters |
| Quarts (liquid) | 32 | Ounces |
| Quarts (liquid) | 2 | Pints |
| Quarts (liquid) | 0.859367 | Quarts (dry) |

| multiply | by | to obtain |
|---|---|---|
| Quarts (liquid) | 64 | Tablespoons |
| Quarts (liquid) | 192 | Teaspoons |
| Quintals | 100 | Kilograms |
| Quintals | 220.462 | Pounds |
| Quires | 25 | Sheets |
| Radians | 57.29578 | Degrees |
| Radians | 3437.747 | Minutes |
| Radians | 0.6366 2 | Quadrants |
| Radians/second | 57.29578 | Degrees/second |
| Radians/second | 9.5493 | Revolutions/minute |
| Radians/second | 0.15915 | Revolutions/second |
| Radians/sec./sec. | 572.96 | Revolutions/min./min. |
| Radians/sec./sec | 0.15915 | Revolutions/sec./sec |
| Reams | 500 | Sheets |
| Revolutions | 360 | Degrees |
| Revolutions | 4 | Quadrants |
| Revolutions | 6.2832 | Radians |
| Revolutions/minute | 6 | Degrees/second |
| Revolutions/minute | 0.1047 | Radians/second |
| Revolutions/minute | 0.01667 | Revolutions/second |
| Revolutions/min./min. | 0.001745 | Radians/sec./sec. |
| Revolutions/min./min | 0.0002778 | Revolutions/sec./sec. |
| Revolutions/second | 360 | Degrees/second |
| Revolutions/second | 6.283 | Radians/second |
| Revolutions/second | 60 | Revolutions/minute |
| Revolutions/sec./sec. | 6.283 | Radians/sec./sec. |
| Revolutions/sec./sec. | 3,600 | Revolutions/min./min. |

| multiply | by | to obtain |
|---|---|---|
| Seconds (angle) | 0.0002778 | Degrees |
| Seconds (angle) | 0.01667 | Minutes |
| Seconds (angle) | 0.000003087 | Quadrants |
| Seconds (angle) | 0.000004848 | Radians |
| Square centimeters | 197,352.5 | Circular mils |
| Square centimeters | 0.001076 | Square feet |
| Square centimeters | 0.155 | Square inches |
| Square centimeters | 0.0001 | Square meters |
| Square centimeters | 100 | Square millimeters |
| Square feet | 0.0000229 | Acres |
| Square feet | 929.03 | Square centimeters |
| Square feet | 144 | Square inches |
| Square feet | 0.0929 | Square meters |
| Square feet | 0.00000003587 | Square miles |
| Square feet | 0.1111 | Square yards |
| Square inches | 1,273,239 | Circular mils |
| Square inches | 6.4516 | Square centimeters |
| Square inches | 0.006944 | Square feet |
| Square inches | 645.16 | Square millimeters |
| Square inches | 1,000,000 | Square mils |
| Square kilometers | 247.1054 | Acres |
| Square kilometers | 10,763,900 | Square feet |
| Square kilometers | 1,000,000 | Square meters |
| Square kilometers | 0.3861 | Square miles |
| Square kilometers | 1,196,000 | Square yards |
| Square meters | 0.0002471 | Acres |
| Square meters | 10.7639 | Square feet |
| Square meters | 0.0000003861 | Square miles |

| multiply | by | to obtain |
|---|---|---|
| Square meters | 1.19599 | Square yards |
| Square miles | 640 | Acres |
| Square miles | 27,878,400 | Square feet |
| Square miles | 2.58999 | Square kilometers |
| Square miles | 3,097,600 | Square yards |
| Square millimeters | 1,973.5 | Circular mils |
| Square millimeters | 0.01 | Square centimeters |
| Square millimeters | 0.00155 | Square inches |
| Square mils | 1.273 | Circular mils |
| Square yards | 0.0002066 | Acres |
| Square yards | 9 | Square feet |
| Square yards | 1,296 | Square inches |
| Square yards | 0.8361 | Square meters |
| Square yards | 0.0000003228 | Square miles |
| Tablespoons | 0.0667 63 | Cubic Centimeters |
| Tablespoons | 0.0625 | Cups |
| Tablespoons | 0.00390625 | Gallons |
| Tablespoons | 0.014787 | Liters |
| Tablespoons | 0.5 | Ounces |
| Tablespoons | 0.03125 | Pints |
| Tablespoons | 0.015625 | Quarts |
| Tablespoons | 3 | Teaspoons |
| Teaspoons | 4.9289 | Cubic Centimeters |
| Teaspoons | 0.020833 | Cups |
| Teaspoons | 0.001302 | Gallons |
| Teaspoons | 0.00493 | Liters |
| Teaspoons | 0.1666667 | Ounces |

| multiply | by | to obtain |
|---|---|---|
| Teaspoons | 0.01042 | Pints |
| Teaspoons | 0.00521 | Quarts |
| Teaspoons | 0.333333 | Tablespoons |
| Tons (short) | 907.18486 | Kilograms |
| Tons (short) | 32,000 | Ounces |
| Tons (short) | 2,000 | Pounds |
| Tons (short) | 0.89286 | Tons (long) |
| Tons (short) | 0.90718 | Tons (metric) |
| Tons (long) | 1,016.05 | Kilograms |
| Tons (long) | 2240 | Pounds |
| Tons (long) | 1.12 | Tons (short) |
| Tons (metric) | 1,000 | Kilograms |
| Tons (metric) | 2204.62 | Pounds |
| Tons of water/24 hrs. | 1.3349 | Cubic feet/hour |
| Tons of water/24 hrs. | 0.16643 | Gallons/minute |
| Tons of water/24 hrs. | 83.333 | Pounds water/hour |
| Yards | 91.44 | Centimeters |
| Yards | 3 | Feet |
| Yards | 36 | Inches |
| Yards | 0.0009114 | Kilometers |
| Yards | 0.9144 | Meters |
| Yards | 0.0005682 | Miles |
| Yards | 0.0004937 | Miles (nautical) |

**11**

# Glossary

**Many of the from the Sunberg Ferar glossary of product development terms)**

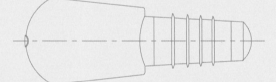

Many of the Glossary entries are drawn from the Sunberg Ferar glossary of product development terms. Extensive explanations or thorough explorations of the following terms may be found in the publications listed in Recommended Reading, pages 264–265

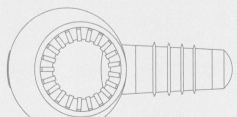

**2D Rendering** A process that industrial designers use to visualize their ideas by putting thoughts on paper with any number of combinations of color markers, pencils, and other tools.

**3D Database** An electronic gathering of three-dimensional product parameters, used for the purpose of product development. The data may be used for testing, prototype development, tooling, and other functions.

**3D Rendering** A process that industrial designers use to visualize their ideas by creating three-dimensional images using computer technology. These renderings also provide 360-degree views of products or concepts.

**Aesthetics** A term that refers to the visual elements or appeal of a design solution.

**Alpha Test** The in-house testing of preproduction products to find and eliminate the most obvious design defects or deficiencies, either in a laboratory setting or in some part of the developing firm's regular operations.

**Amortization** A term related to tooling costs, which refers to the period of time over which the cost of building tools must be spread (amortization is a factor of a percentage of profits and time).

**Appearance Model** A physical example of a new product concept. In the purest sense, a model is a solid rendering of a product. Unlike a typical prototype, a model is required to be neither functional nor representative in materials. A prototype may serve as an appearance model. However, an appearance model is not necessarily a prototype.

**Attributes** Inherent characteristics or qualities of a design ascribed in words.

**Barrier to Entry** A financial, technical, material, or other obstacle that must be overcome to do business in a given business category.

**Benchmarking** A process of studying successful competitors (or organizations in general) and selecting the best of their actions or standards. In the new product program, it means finding the best development process methods and the best process times to market, and then setting out to achieve them.

**Beta Test** An external test of preproduction products. The purpose is to test the product before sale to the general market for all functions in a breadth of field situations, to find system faults that are more likely to appear in actual use than in the firm's more controlled in-house tests.

**Bill of Materials (BOM)** A listing of all subassemblies, intermediate parts, and raw materials that go into a parent assembly. This document outlines the quantity of each part required to make an assembly.

**Blue Sky** Unconstrained by practical concerns for the purpose of inspiring ideas or pushing the envelope.

**Brainstorming** Group methods of problem-solving used in product concept generation; there are many modifications in format of use, each variation with its own name.

**Brand** A name, term, design, symbol, or any other feature that identifies one seller's goods or services as distinct from those of other sellers. The legal term for brand is trademark. A brand may identify one item, a family of items, or all items of that seller.

**Breadboard** A proof-of-concept modeling technique that represents how a product will work, but not how a product will look.

**Business Plan** A written blueprint for a company, describing objectives, personnel structure, financial models, capabilities, etc. This document is typically used to solicit investment or partnership.

**Buyer** The purchaser of a product, whether or not they will be the ultimate user. Especially in business-to-business markets, a purchasing agent may contract for the actual purchase of goods or services yet never benefit from them.

**Cannibalization** When the demand for a new product erodes demand for (sales of) a current product the firm offers.

**Champion** A person who takes an inordinate interest in seeing that a particular process or product is fully developed and marketed. The role varies from situations calling for little more than stimulating awareness of the opportunity to extreme cases where the champion tries to force a project past entrenched internal resistance in the company policy or from objecting parties.

**Commodity** An item that is part of a mature market, which is not expected to grow and where all products exhibit sameness in style, features, or other aspects.

**Competitive Advantage** An advantage over competitors' products, achieved through functional benefit or end-user emotional benefit.

**Computer Assisted Design (CAD)** A technology that allows designers and engineers to use computers for their design work. Also termed CAID (industrial design) and CAE (engineering).

**Concept** A clearly written and possibly visual description of a new product idea that includes its primary features and consumer benefits.

**Concept Development** A broad term used to describe the set of activities associated with transforming generated concepts into products. Activities include research, prototype build, and testing.

**Concept Generation** The act by which new concepts, or product ideas, are generated. Sometimes also called idea generation or ideation.

**Concept Testing** The process by which a concept statement is presented to consumers for their reactions. These reactions can either be used to permit the developer to estimate the sales value of the concept or to make changes to the concept to enhance its potential sales value.

**Concurrent Engineering** When product design and manufacturing process development occur concurrently or simultaneously rather than sequentially. Also called simultaneous engineering.

**Conjoint Analysis** A quantitative market-research technique that determines how consumers make trade-offs between a small number of different features or benefits.

**Consumer** One who purchases or uses your firm's products or services; not to be confused with customer or end user.

**Consumer Market** The purchasing of goods and services by individuals for household use (rather than for use in business settings). Individual decision makers generally make consumer purchases for either themselves or others in the family.

**Convergent Thinking** A technique generally performed in the initial phase of idea generation to help funnel the high volume of ideas created through divergent thinking into a small group or single idea on which more effort will be focused.

**Cost-Effective** A term used to describe the general awareness during the product development process toward meeting cost constraints. Factors considered are related to the bill of materials and cost of design and development, tooling, and manufacturing.

**Cost of Goods Sold (COGS)** The direct costs associated with producing a product.

**Criteria** Statements of standards used by gatekeepers at each gate, related to all organizational functions. The criteria must be achieved or surpassed for product development projects to continue in development. Taken together, these criteria reflect a business unit's new product strategy.

**Customer** The most generic and all-encompassing term for a firm's targets. The term is used in either the business-to-business or the household context, and may refer to the firm's current customers, competitors' customers, or current nonpurchasers with similar needs or demographic characteristics. The term does not differentiate between whether the person is a buyer or a user target. Only a fraction of consumers will become customers. *See* End User.

**Customer Value Added Ratio** The ratio of WWPF (worth what paid for) for your products to WWPF for your competitors' products. A ratio above 100 percent indicates superior value compared to your competitors.

**Deliverable** The completed end result or outcome of a series of tasks.

**Design** Devise or create for a specific purpose or function.

**Design Brief (Design Criteria)** A formal document containing standards agreed upon by key stakeholders. It includes prioritized wants and needs, and is used to measure the success of designs relative to the project goals.

**Design for Manufacturing and Assembly (DFMA)** A method of designing parts and assemblies that makes them easier to put together. Involves techniques such as top-down Z-Axiz assembly; part consolidation; and elimination of fasteners and asymmetrical parts that only go together one way, etc.

**Distribution** The method partners use to get the product (or service) from where it is produced to where the end user can buy it.

**Divergent Thinking** A technique performed early in the initial phase of idea generation, which expands thinking processes to record and recall a high volume of new or interesting ideas.

**Early Adopters** Customers who, relying on their own intuition and vision, buy into new product concepts very early in the life cycle. Also refers to organizational entities that are willing to try out new processes rather than just maintaining the old.

**Eco-Design** Design intended to reduce the consumption of resources in manufacturing products that are designed for recycling.

**End User** The end user is the customer, consumer, or person who uses a product. End user is a more precise term than consumer, because a consumer may be a retail buyer, distributor, etc., who may never actually use the product. Like the consumer, the end user may not always be the only customer.

**Engineering Design** A function in the product-creation process where a good is configured and specific form is decided.

**Engineering Model** The combination of hardware and software intended to demonstrate the simulated functioning of the intended product as currently designed.

**Entrepreneur** An individual who organizes, manages, and assumes the risks of a business enterprise.

**Ergonomic Study** A methodical evaluation of ergonomics, components, and attributes. An ergonomic study can consist of an evaluation of an existing product, evaluation of a new product or prototype, or the creation of a testing tool and a test plan that enables ergonomic options to be explored.

**Ethnography** A descriptive, qualitative market research methodology for studying the customer in relation to his or her environment. Researchers spend time in the field observing customers and their environment to acquire a deep understanding of their lifestyles or cultures as a basis for better understanding their needs and problems.

**Factory Cost** The cost of producing the product in the production location, including materials, labor, and overhead.

**Failure Rate** The percentage of a firm's new products that make it to full market commercialization but that fail to achieve the objectives set for them.

**Feature** The solution to a consumer need or problem. Features are the way benefits are provided to consumers. Usually any one of several different features may be chosen to meet a customer need. For example, a carrying case with a shoulder strap is a feature that allows a laptop computer to be carried easily.

**Feature Creep** The tendency for designers or engineers to add more capability, functions, and features to a product as development proceeds than were originally intended. These additions frequently cause schedules to slip, development costs to increase, and product costs to rise.

**Field Testing** Product use testing with users from the target market.

**File Transfer Protocol (FTP)** Allows a computer to send or receive files over the internet. FTP works by means of a client-server architecture; the user runs client software to connect to a server on the Internet. The FTP server allows the user to download and upload files.

**Finite Element Analysis (FEA)** A mathematical technique for analyzing stress or failure points. The structure or system is divided into substructures called "finite elements." The finite elements and their interrelationships can be evaluated for the purpose of validation or redesign. FEA software is available for most popular CAD packages.

**First-To-Market** The first product that creates a new product category or a substantial subdivision of a category.

**Focus Groups** A qualitative market research technique where eight to twelve market participants are gathered in one room for a discussion under the leadership of a trained moderator. The discussion focuses on a consumer problem, product, or potential solution to a problem.

**Form Study** A model created for the purpose of evaluating the shape or feel of a product. Form studies typically do not reflect manufacturing or functional reality.

**Gantt Chart** A horizontal bar chart used in project scheduling that shows the start date, end date, and duration of tasks within the project. It is sometimes used in conjunction with a network diagram.

**Gate** The decision point; often a meeting at which a management decision is made to allow the product development project to proceed to the next stage, or to recycle back into the current stage to better complete some of the tasks, or to terminate. The number of gates varies by company.

**Gatekeeper** A gatekeeper is anyone who will touch, see, or otherwise interact with a product throughout its development.

**GUI "gooey"** Graphic User Interface is a touch-screen or control panel that provides graphic direction of selection options.

**Haptics** Relating to or based on the sense of touch.

**Human Factors/Ergonomic Study** The study of the dimensions, range of motion, and physical limitations of the human form, and the application of those factors to design.

**Human Interface Design** The design of the product elements the user physically interacts with. Typically, these elements are the controls.

**Idea Generation (Ideation)** All the activities and processes that lead to creating new product or service ideas that may warrant development.

**Incremental Improvement** A small change made to an existing product that serves to keep the product fresh in the eyes of customers.

**Industrial Design (ID)** The professional service of creating and developing concepts and specifications that optimize the function, value, and appearance of products and systems for the mutual benefit of both user and manufacturer (Industrial Design Society of America).

**Innovation** A new idea, method, or device. The act of creating a new product or process. The act includes invention as well as the work required to bring an idea or concept into its final form.

**Integral Architecture** A product architecture in which most or all of the functional elements map into a single or very small number of chunks. It is difficult to subdivide an integrally designed product into partially functioning components.

**Interdisciplinary** Involving two or more academic, scientific, or artistic disciplines. Typically referring to interdisciplinary collaboration between designers, engineers, researchers, or marketers.

**Intradisciplinary** Involving two or more individuals from like disciplines. Typically referring to intradisciplinary collaboration between design professionals.

**Invention** The act of striving to create a new method, solution, process, or device by testing or studying an idea or groups of ideas. Invention may or may not result in a desired outcome; a solution may not be reached to the satisfaction of all participants.

**ISO-9000** A set of five auditable standards of the International Standards Organization that establishes the role of a quality system in a company and that is used to assess whether the company can be certified as compliant to the standards. ISO-9001 deals specifically with new products.

**Lead Users** Users for whom finding a solution to a need is so important that they have modified a current product or invented a new one to solve the need themselves because they have not found a supplier who can solve it for them. When these consumers' needs are portents of needs that the center of the market will have in the future, their solutions are new product opportunities.

**Lean Manufacturing** Reducing the time from customer order to manufacturing and delivering products by eliminating wasted time and effort in the production stream.

**Left-Brained** A term used to describe individuals or ways of thinking that are closely aligned with analytical and business practices. Terms such as Return on Investment (ROI), process, bottom-line, and timing are associated with left-brained thinking.

**Line Extension** A form of derivative product that adds or modifies features without significantly changing the price.

**Manufacturability** The extent to which a new product can be easily and effectively manufactured at minimum cost and with maximum reliability.

**Manufacturing Design** Determining the manufacturing process that will be used to make a new product.

**Market Conditions** The characteristics of the market into which a new product will be placed, including the number of competing products, level of competitiveness, and growth rate.

**Market Driven** When the marketplace is allowed to direct a firm's product innovation efforts.

**Market Segmentation** The act of dividing an overall market into groups of consumers with similar needs, where each of the groups differs from others in the market in some way.

**Market Share** A company's sales in a product area given as a percent of the total market sales in that area.

**Market Testing** The product development stage during which the new product and its marketing plan are tested together. A market test simulates the eventual marketing mix and takes many different forms, only one of which bears the name test market.

**Mass Customization** The use of flexible computer-aided manufacturing systems to produce custom output. Those systems combine the low unit costs of mass production processes with the flexibility of individual customization.

**Maturity Stage** The third stage of the product life cycle. This is the stage where sales begin to level due to heavy competition and market saturation, alternative product options, or changing buyer or user preferences.

**Metrics** A set of measurements to track process, action, or relationship, allowing one to measure the impact of process changes or improvement over time.

**Modular Architecture** A type of product architecture in which each functional element maps into its own physical chunk. Different chunks perform different functions.

**Multifunctional Team** A group of individuals brought together from more than one functional area of a business to work on a problem or process that requires the knowledge, training, and capabilities across the areas to complete the work.

**Net Present Value (NPV)** A method used in comparably evaluating investments in very dissimilar projects by discounting the current and projected future cash inflows and outflows back to the present value based on the discount rate, or cost of capital, of the firm.

**Observational Research** A qualitative method of directly observing how end users interact with products; also relates to the study of man in his environment and the study of man interacting with products in his environment.

**Operations** A term that includes manufacturing but is much broader, usually including procurement, physical distribution, and, for services, management of the offices or other areas where the services are provided.

**Operator's Manual** The written instructions to the users of a product or process. These may be intended for the ultimate customer or for the use of the manufacturing operation.

**Payback** The time, usually in years, from some point in the development process until the commercialized product or service has recovered its costs of development and marketing. While some firms take the point of full-scale market introduction of a new product as the starting point, others begin the clock at the start of development expense.

**Performance Indicators** Criteria with which the performance of a new product in the market can be evaluated.

**Platform Product** The design and components that are shared by a set of products in a product family. From this platform, numerous derivative products can be designed.

**Preliminary Bill of Materials (PBOM)** A forecasted listing of all the subassemblies, intermediate parts, raw materials, engineering design, tool design, and customer inputs that are expected to go into a parent assembly, with an indication of anticipated quantities of each component.

**Preproduction Unit** A product that looks and acts like the intended final product, but is made by hand or in pilot facilities rather than by the final production process.

**Process Map** A workflow diagram that uses an x-axis for process time and a y-axis that shows participants and tasks.

**Process Managers** The operational managers responsible for ensuring the orderly and timely flow of ideas and projects through the process.

**Product Architecture** The way in which functional elements are assigned to the physical chunks of a product and the way those physical chunks interact.

**Product Definition** Defines the product, including the target market, product concept, benefits to be delivered, positioning strategy, price points, product requirements, and design specifications.

**Product Development** The overall process of strategy, organization, concept generation, product- and marketing-plan creation and evaluation, and commercialization of a new product.

**Product Development Process** A disciplined and defined set of tasks and steps that describe the means by which a company repetitively converts embryonic ideas into salable products or services.

**Product Development Strategy** The strategy that guides the product innovation program.

**Product Development Team** A multifunctional group of individuals chartered to plan and execute a new product development project.

**Product Failure** A product development project that does not meet the objective of its developers.

**Product Family** A set of products derived from a common product platform. Members of a product family normally have many common parts and assemblies.

**Product Interfaces** Internal and external interfaces impacting the product development effort, including the nature of the interface, action required, and timing.

**Product Life Cycle** The four stages a new product is thought to go through from birth to death: introduction, growth, maturity, and decline. Controversy surrounds whether products go through this cycle in any predictable way.

**Product Line** A group of products marketed by an organization to one general market. The products have some characteristics, customers, and uses in common, and may also share technologies, distribution channels, prices, services, and other elements of the marketing mix.

**Product Manager** The person assigned responsibility for overseeing all the various activities that concern a particular product; sometimes called a brand manager in consumer packaged-goods firms.

**Product Plan** A detailed summary of the key elements involved in a new product development effort, such as product description, schedule, resources, financial estimations, and interface management plan.

**Product Platforms** Underlying structures or basic architectures that are common across a group of products, or that will be the basis of a series of products commercialized over a number of years.

**Project Management** Both a process and set of tools and techniques concerned with defining the project's goal, planning all the work to reach the goal, leading the project and support teams, monitoring progress, and seeing to it that the project is completed in a satisfactory way.

**Protocol** A statement of the attributes a new product is expected to have. A protocol is prepared prior to assigning the project to the technical development team. The benefits statement is agreed to by all parties involved in the project.

**Prototype** A physical model of the new product concept. Depending upon the purpose, prototypes may be nonworking, functionally working, or both functionally and aesthetically complete. A prototype is typically created to test or prove a functional aspect of a product. A prototype may serve as an appearance model. However, an appearance model is not necessarily a prototype.

**Psychographics** Characteristics of consumers that measure their attitudes, interests, opinions, and lifestyles rather than their demographics.

**Q-Sorts** A process for sorting and ranking complex issues.

**Qualitative Market Research** Consumer research conducted with a very small number of consumers, either in groups or as individuals. Results are not necessarily representative of consumers in general, nor are they projected. Frequently used to gather initial consumer needs and obtain initial reactions to ideas and concepts.

**Quality Function Deployment (QFD)** A structured method employing matrix analysis for linking what the market requires to how it will be accomplished in the development effort. This method is most valuable during the stage of development when a multifunctional team agrees on how customer needs relate to product specifications and features. By explicitly linking these aspects of product design, QFD limits the chance of omitting important design characteristics or interactions across design characteristics. QFD is also an important mechanism in promoting multifunctional teamwork.

**Quantitative Market Research**
Consumer research, often surveys, conducted with a large enough sample (200 respondents) of consumers to produce statistically reliable results that can be used to project outcomes to the general consumer population. Used to determine importance levels of different customer needs; performance ratings of, and satisfaction with, current products; and probability of trial, repurchase rate, and product preferences. These techniques are used to reduce the uncertainty associated with many other aspects of product development.

**Rapid Prototyping** Any of a variety of processes that avoids tooling time in producing prototypes or prototype parts and therefore allows (generally nonfunctioning) prototypes to be produced within hours or days rather than weeks. These prototypes are frequently used to quickly test the product's technical feasibility or consumer interest.

**Reposition** To change the product positioning due to failure of the original positioning, or react to changes in the marketplace. Most frequently, repositioning is accomplished solely through changing the marketing mix.

**Resource Plan** A detailed summary of all of resources required to complete product development, including personnel, equipment, time, and finances.

**Return on Investment (ROI)** A standard measure of project profitability, ROI is the discounted profits over the life of the project expressed as a percentage of initial investment.

**Right-Brained** A term used to describe individuals or ways of thinking that are closely aligned with creativity, passion, and intuition. Terms such as design, aesthetics, and innovation are associated with right-brained thinking.

**Segmentation** The process of dividing a large and heterogeneous market into more homogeneous subgroups. Each subgroup, or segment, holds similar views about the product, and purchases and uses the product in similar ways.

**Semantics** The study of language meaning.

**Semiotics** A philosophical theory of signs and symbols that deals with their function in both artificially constructed and natural languages, comprising syntactics, semantics, and pragmatics.

**Senior Management** The level of executive or operational management above the product-development team, which has approval authority or controls resources important to the development effort.

**Services** Products, such as airline flights or insurance policies, that are intangible—at least substantially so. If totally intangible, they are exchanged directly from producer to user, cannot be transported or stored, and are instantly perishable. Service delivery usually involves customer participation in some important way, and cannot be sold in the sense of ownership transfer.

**Specification** A detailed description of the features and performance character-istics of a product. For example, a lap-top computer's specification may read as a 90-megahertz Pentium, with 16 mega-bytes of ram and 720 megabytes of hard disk space, 3.5 hours of battery life, weighing 4.5 pounds, with an active-matrix 256-color screen.

**Spreadsheet** A table of values arranged in rows and columns. Each value may have a predefined relationship to the other values, so if you change one value, you may need to change other values as well.

**Stakeholder** Typically, stakeholders are those who are involved in the product life-cycle decision within an organization. These people include marketing, manu-facturing, finance, advertising, and government regulators, among others, who have a stake in the outcome of a product. Stakeholders may include recy-clers and consumers. See also *Design Brief* and *Gatekeeper*.

**Steel-Safe Change** When a steel manu-facturing tool is being made (such as an injection molding tool), material is removed to create the negative cavity. It is very difficult to add metal back to the tool once it has been removed. A steel-safe change to a tool means the usabil-ity of the tool is not in jeopardy.

**Storyboard** Related to product develop-ment, storyboards are grouped render-ings or images of products or concepts that are used to demonstrate an approach or philosophy. The "board" typically refers to the foam core on which these images are mounted.

**Styling** The form given to an object.

**Subassembly** A collection of compo-nents that can be put together as a sin-gle assembly to be inserted into a larger assembly or final product. Often the subassembly is tested for its ability to meet explicit specifications before inclusion in the larger product.

**Target Market** The group of consumers or potential customers selected for mar-keting; a market segment of consumers.

**Team** The group of persons who partici-pate or manage participation in the product development project. Frequently, each team member represents a func-tion, department, or specialty; together, they provide the full set of capabilities needed to complete the project, in which case they are referred to as a multifunc-tional team.

**Test Market(s)** One or more limited geographic regions into which a new product is launched in a very controlled manner. Consumer response to the product and its launch is measured to monitor and gauge a product's viability. When multiple geographies are used in the test, different advertising or pricing policies may be tested and the results compared.

**Think Tank** Environments created by management to generate new ideas, or approaches to solving organizational problems.

**Thumbnail** The most minimal form of sketching, usually using pencils, to represent a product idea. Also known as a napkin sketch.

**Time to Market** The length of time it takes to develop a new product from an early idea for a new product to initial market sales. Precise definitions of the start and end points vary from one company to another and may vary from one project to another within the company.

**Tone** The feeling, emotion, or attitude most associated with using a product. The appropriate tone is important to include in new product concepts and advertising for consumers.

**Tooling** For plastic injection molding, the solid stock of materials that serve as molds for products. Tools are synonymous with molds, although tools typically have a longer life cycle than molds that are made of softer materials. Tools may be developed for production parts or prototype parts. Prototype tools are usually built of less permanent material (such as aluminum) and are used to create a limited number of parts to be used for testing of engineering and manufacturing elements. Other types of tooling are used for manufacturing requiring stamping, roll-forming, extruding, die-casting, and the like.

**Tooling Lead Time** The time required to create tools for the production of product parts. The lead time may be many months for large tools. Tooling lead time is a critical factor to consider when determining the viability of a projected product launch time.

**Tweak** A term used to describe a minor refinement or adjustment.

**Universal Design** Design intended to meet the needs of a broad customer base, including those with physical disabilities.

**User** Any person who uses a product or service to solve a problem or obtain a benefit, whether or not they purchase it. Users may consume a product, as in the case of a person using shampoo to clean his or her hair, or eating a potato chip to assuage hunger between meals. Users may not directly consume a product, but may interact with it over a longer period, such as in a family owning a car, with multiple family members using it for many purposes over a number of years. Products also are employed in the production of other products or services, where the users may be the manufacturing personnel who operate the equipment.

**Value-Added** The act or process by which tangible product features or intangible service attributes are bundled, combined, or packaged with other features and attributes to create a competitive advantage, reposition a product, or increase sales.

**Vertical Integration** The expansion of a business or capabilities by acquiring or developing businesses with those capabilities. In forward vertical integration, manufacturers might acquire or develop wholesale and retail activities. In backward vertical integration, retailers might develop their own wholesale or manufacturing capabilities.

**Voice of the Customer (VOC)** A process for eliciting needs from consumers that uses structured in-depth interviews to lead interviewees through a series of situations they have experienced in which they found solutions to the set of problems being investigated. Needs are obtained through indirect questioning by coming to understand how the consumers found ways to meet their needs, and more importantly, why they chose the particular solutions they found.

## RECOMMENDED READING

*Bodyspace: Anthropometry, Ergonomics and the Design of the Work*
Stephen Pheasant
CRC Press, 2nd ed., 1996

*Brand Leadership*
David A. Aacker
Free Press, 2000

*Competitive Advantage: Creating and Sustaining Superior Performance*
Michel E. Porter
Free Press, 1985

*Cradle to Cradle: Remaking the Way We Make Things*
William McDonough and Michael Braungart
North Point Press, 2002

*Designing Pleasurable Products: An Introduction to the New Human Factors*
Patrick Jordan
CRC Press, 2002

*Dictionary of Business Terms*
Jack P. Friedman
Barron's Educational Series, 3rd ed., 2000

*Digital Design and Manufacturing: CAD/CAM Applications in Architecture and Design*
Daniel Schodek, Martin Bechthold, James Kimo Griggs, Kenneth Kao, and Marco Steinberg
Wiley, 2005

*Eco Design: The Sourcebook*
Alastair Fuad-Luke
Chronicle Books, 2002

*Ergonomics for Beginners: A Quick Reference Guide*
Jan Dul and Bernard Weerdmeester
CRC Press, 2001

*Forms, Folds, and Sizes: All the Details Graphic Designers Need to Know But Can Never Find*
Poppy Evans
Rockport Publishers, 2004

*Human Factors in Engineering and Design*
Mark S. Sanders and Ernest J. McCormick
McGraw-Hill Science, 1993

*Industrial Design Materials and Manufacturing: Materials and Manufacturing*
Jim Lesko
Wiley, 1999

*Management and Organisational Behaviour*
Laurie Mullins
Financial Times, Pitman Publishing, 7th ed., 2004

*Managing Brand Equity*
David A. Aacker
Free Press, 1991

*Marketing Management*
Philip Kotler
Prentice-Hall, 11th ed., 2002

*The Measure of Man and Woman: Human Factors in Design*
Alvin R. Tilley and Henry Dreyfuss Associates
Wiley, 2001

*Power In and Around Organizations (The Theory of Management Policy Series)*
Henry Mintzberg
Prentice-Hall, 1983

*Product Design and Development*
Karl T. Ulrich and Steven D. Eppinger
McGraw-Hill, 2003

*Product Strategy and Management*
Michael Baker and Susan Hart
Prentice-Hall, 1999

*The Strategic Management of Organizations*
Adrian Haberberg and Alison Rieple
Pearson Education Limited, 2001

*Strategy: Process, Content, Context—An International Perspective*
Bob de Wit and Ron Meyer
International Thompson Business Press, 1999

*Strategy Safari: A Guided Tour Through the Wilds of Strategic Management*
Henry Mintzberg, Bruce Ahlstrand, and Joseph Lampel
Free Press, 1998

*Universal Principles of Design: 100 Ways to Enhance Usability, Influence Perception, Increase Appeal, Make Better Design Decisions, and Teach Through Design*
William Lidwell, Kritina Holden, and Jill Butler
Rockport Publishers, 2003

# INDEX

## A

abrasive finishing, 58
AC induction motors, 99
AC/DC motors, 99
adhesive bonding, 51, 59
alloy steel, 81
alternating current, 94
aluminum, 82
amps, 90–91
annealing, 72
anodizing, 58
anthropometrics, 118–127, 169
arc welding, 64
arm and elbow, anatomy of, 162
arrowheads, 189, 192, 193
automated glass forming, 71

## B

baseline dimensioning, 191
batteries, 93–94
beryllium, 82
bisque, 67
bisque firing, 67
blow and blow process, 71
blow molding, 36
Board of Directors, 14
bone dry, 67
brands
  basic concepts, 20–21
  importance of, 10 20
brazing, 59
business environment, 10–21
brands, 18–21
business, categories of, 13–14
competitive forces, 16–18
marketing, 13–14
strategy, 15–16

## C

carbon steel, 81
cast iron, 81
casting, 38, 54
ceramics
  defined, 80
  overview, 67
  terminology, 67–70
chain dimensioning, 191
chamfers, 196
Chief Executive Officer (CEO), 14
child development, 138–145
Child Safety Protection
  Act, 135–137
children, interest and abilities of
  6-year-olds, 142–143
  7-year-olds, 143
  8-year-olds, 144
  infants, 138–139
  preschoolers, 141–142
  toddlers, 140–141

children, motor development
  milestones, 146
circle
  area of, 203
  circumference of, 203
  hollow, 203
  sector of, 203
  segment of, 204
circuits, 90–91
clay body, 67
coating, 58, 72
collar, 60
color systems, 218
competition, 16–18
component manufacturing, 172
composites, 80
compression molding, 38
computer workstations,
  design of, 155–161
computerized numerical
  control (CNC), 74
concept development, 26
cone
  area of, 206
  volume of, 206
contacts, 93
container forming, 72
contour roll forming, 55
control panel layout, 128–131
conversions
  area, 224
  capacity, 223–224
  length, 222–223
  miscellaneous, 228–246
  standard inch fractions to
    decimals, 226–227
  type, 227
  volume, 225
  weight, 220–246
copper, 82
copyright, 106
cylinder
  area of, 205
  volume of, 205

## D

data exchange, of digital
  files, 210–211
decimals, 197
decorating, 72
deep draw (stamping), 56
design
  business context of, 10–21
  of computer workstations,
    155–161
  of hand tools, 162–168
  research, 26
  sustainable, 170–175
  of toys, 138–145

design documentation, 176–211
  3D modeling, 208–211
  dimensioning, 188–200
  example of complete, 207
  layout, 181–192
  line weights and styles,
    182–185
  management of, 211
  overview, 176
  scale drawings, 198–200
  standard formats and sizes,
    178–180
  title box, 200
  volume calculations, 201–207
detail views, 186
die casting, 55
Diffrient, Niels, 116, 153, 160
digital files, 208
dimension lines, 189, 193
dimensioning, 188–200
  fractions and decimals, 197
  to intersection with a radius
    edge, 197
  standards for, 193–196
  systems for, 194
  techniques, 191–192
diodes, 101–102
disposal, 173
distribution, 172
distributors, 17–18
document management, 211
drafting formats, 178–180

## E

earthenware, 80
ecodesign practices, 174
  See also sustainable design
elastomers, 80
electric motors, 98–99
electrical discharge machining
  (EDM), 33
electrical sockets, types of worldwide, 103
electromagnets, 97–98
electronics, 88–103
  alternating current, 94
  batteries, 93–94
  contacts, 93
  diodes, 101–102
  electrical circuit, basic, 90–91
  electromagnets, 97–98
  multielement circuits, 92
  overview, 88
  power distribution and
    safety, 95–96
  rectifiers, 101–102
  switches, 93
  transformers, 100–101
electroplating, 46–47
electrostatic painting, 58

**Carla J. Blackman** is a principal at Design Interface Inc., an award-winning design consultancy specializing in industrial, graphic, package, and Web design. She has taught ergonomics at the Cleveland Institute of Art since 1987 and continues to mentor students in her role as student advisor for IDSA. She judged the 2002 IDEA Awards sponsored by *Business Week* and IDSA.

**Darrell E. Covert** has developed a wide range of patented products while working for Fortune 500 companies and some of the world's premier consultancies. He holds a BS in industrial design and has more than twenty-six years of design experience. He has had the privilege of heading design offices in both the U.S. and Europe and holds more than thirty-nine U.S. patents and fourteen international patents. In 1999, he received the "Distinguished Corporate Inventors Award" from Goodyear. He is well versed in the strategic advantage and use of patents as a vital competitive design skill. His clients include Goodyear, General Motors, Chrysler, IBM, Tupperware, Fisher Price, Motorola, Boeing, Gates Lear, Air Bus, Pininfarina, Ital Design, Bertone, Porsche, and Lotus.

**Daniel F. Cuffaro** is the chairman of the Industrial Design Department at the Cleveland Institute of Art and an associate professor. He has written several articles for Innovation Magazine and *the Design Management Institute Journal* and has lectured in Mexico, the United Kingdom, and around the United States. He is the former director of design at Altitude, Inc., and continues to consult in the areas of industrial design, strategy, economic development, and urban planning.

**David Laituri** is the director of product development at Brookstone. Prior to Brookstone, David held the position of director of industrial and user interface design with Polaroid Corporation. During his twenty-year career, his client list has included Apple Computer, General Motors, Hewlett Packard, Nike, Oxo, Hasbro, and Proctor & Gamble. David holds an undergraduate degree in industrial design from Ohio State University as well as an MBA from the University of Westminster in London.

(continued)

(continued from page 271)

**Amy Nehez-Cuffaro** is a freelance art director/graphic designer and teaches at the Cleveland Institute of Art.

**Douglas Paige** is an associate professor of industrial design at the Cleveland Institute of Art. He has lectured at design schools and business organizations and been a guest diploma juror at schools in France and Germany. He is a former senior designer and manager with Thomson Consumer Electronics Inc., and continues an active practice in Industrial Design.

**Lawrence M. Sears** received a BS in Electrical Engineering from Case Institute of Technology. He is founder and Director of Technology of Hexagram, Inc., one of the leading providers of products for automatically reading utility meters. Mr. Sears holds or coholds nineteen patents relating to a variety of industrial, commercial, and communications products. Mr. Sears is also a lecturer in the Department of Electrical Engineering and Computer Science at Case Western Reserve University, in Cleveland.

---

## Acknowledgments

The authors would like to thank a number of people without whom this work would have been impossible. These include Brian Matt, Heather Andrus, Eduardo Milrud, Frank Cuffaro, Dave Jetter, Larry Nagode, and William Brouillard. We would especially like to thank the team at Rockport Publishers, especially Kristin Ellison, Rochelle Bourgault, Regina Grenier, and Sylvia McArdle, for their guidance, hard work, and expertise.